Cold Plasma

Cold Plasma

Characteristics and Applications in Medicine

Selected Articles Published by MDPI

MDPI • Basel • Beijing • Wuhan • Barcelona • Belgrade • Manchester • Tokyo • Cluj • Tianjin

Contents

About the Editor

Mounir Laroussi, Dr., received his Ph.D. in Electrical Engineering from the University of Tennessee, Knoxville. He is currently Professor at the Electrical & Computer Engineering Department of Old Dominion University (ODU) and Director of ODU's Plasma Engineering & Medicine Institute (PEMI).

Dr. Laroussi's research interests are in the physics and applications of non-equilibrium gaseous discharges, including the biomedical applications of low-temperature plasma (LTP). He has designed and developed numerous novel LTP devices, such as resistive barrier discharge (RBD) and the plasma pencil. He is the co-discoverer of guided ionization waves in low-temperature plasma jets. Dr. Laroussi is also widely known for conducting the first pioneering experiments on the use of low-temperature atmospheric pressure plasmas for biomedical applications and for highly contributing to the establishment of the interdisciplinary field of plasma medicine. For his scientific achievements in the field of low-temperature plasmas and their biomedical applications, he was elevated to the grade of fellow by IEEE in 2009 and has been awarded the 2012 IEEE-NPSS Merit Award in addition to other prestigious awards.

Dr. Laroussi is the author or co-author of two books and more than 200 papers in refereed journals and conference proceedings, and he holds seven patents in the field of plasma devices and their applications. He served as the general chair of the 2010 IEEE International Conference on Plasma Science (ICOPS) and is the co-founder of the International Workshop of Plasma for Cancer Treatment (IWPCT). Dr. Laroussi's research has been featured in various well-known magazines, such as National Geographic, Physics Today, and Scientific American. His work has also been featured in numerous science and technology documentaries.

Dr. Mounir Laroussi (Photograph by Glen McClure)

Preface to "Cold Plasma"

Low Temperature Plasma Sources: Characterization and Biomedical Applications

Mounir Laroussi

For many decades non-equilibrium plasmas (NEPs) that can be generated at atmospheric pressure have played important roles in various material and surface processing applications. Although there are many methods to generate NEPs, one of the simplest and most practical ways is to use the dielectric barrier discharge (DBD) configuration. This discharge uses a dielectric to cover at least one of two electrodes. The plasma generated in the gap between electrodes is generally filamentary, but under some conditions can be uniformly diffuse. Extensive research work has been done on DBD, and one of its earliest applications was to generate ozone for the cleaning of water supplies [1–4]. DBD has also gained widespread use in biomedical applications since the mid 1990s, when it was demonstrated that the plasma produced by DBD possesses strong germicidal properties [5]. However, because the plasma is confined to the gap between electrodes, the use of the conventional DBD setup in biomedical applications has remained limited. This situation changed when investigators reported that with proper design the plasma can be "blown" out of the discharge gap and into the ambient air [6,7]. This development has opened up all kind of possibilities to use this plasma arrangement for medical applications. This is because the plasma can be made available completely out of the ignition region and launched via a nozzle into ambient air. Therefore, it can be aimed at a specific location (such as a wound) and applied for a certain length of time to achieve a biological outcome. Since all this can be done at atmospheric pressure and in ambient air it has become possible to treat actual patients with such plasma generation schemes. These devices have come to be known as non-equilibrium atmospheric pressure plasma jets (N-APPJ).

The first applications of N-APPJs were in material processing. Using various operating conditions and gases, they were found to increase the wettability of polypropylene (PP) and polyethylene terephthalate (PET) films [8]; to degrade aromatic rings of dies such methyl violet [9]; to etch silicon and Si (100); to ash photoresist at a rate greater than 1.2 µm/min [10]; to deposit silicon dioxide, SiO_2, and films on various substrates at deposition rates greater than 10 nm/s; and so on. However, the biomedical applications of N-APPJs only surged after the first "bio-tolerant" plasma jets were reported in the mid-2000s [6,7,11]. Today these plasma jets and other plasma sources are being extensively researched for medical applications ranging from wound healing to dentistry and cancer therapy [12–27].

Around 2005, investigators at Old Dominion University, USA, and the University of Wuppertal, Germany, independently discovered that the plasma plumes of N-APPJs were in fact not continuous but made of fast-propagating discrete small volumes of plasma (known as "plasma bullets") [28,29]. This has led to numerous experimental and modeling works aimed at elucidating the mechanisms of ignition and propagation of N-APPJs [30–34]. Recently it was well established that these jets are enabled by guided ionization waves where photoionization and the electric field at the head of the ionization front play important roles [35]. The magnitude of the electric field was measured by several investigators and was found to be in the 10–30 kV/cm range [36–38].

Various power-driving methods have been used to ignite and sustain N-APPJs. These include DC, pulsed DC, RF, and microwave power [7]. Because they provide interesting reactive chemistry, N-APPJs play an ever-increasing role in biomedicine. Reactive oxygen species (ROSs) and reactive nitrogen species (RNSs) such as O, OH, O_2^-, $^1O_2$, H_2O_2, NO, and NO_2, which are generated by these plasma jets, have been shown to play a central role in their interactions with liquids and soft matter, including cells and tissues [39–42]. Based on these results it has been concluded that the biological effects of these plasmas are mostly mediated by the ROSs and RNSs they produce. These reactive

"

molecules (radical and non-radical) can oxidize membranes' lipids and proteins and can trigger and/or
modulate cell signaling. Depending on the type and concentration of ROS and RNS, proliferation or
destruction of cells can occur. Many investigators have reported that low temperature plasma can be
tailored to induce apoptosis in cancer cells without causing damage to healthy cells [43–45]. It is also
suspected that the high electric field at the head of plasma plumes can cause electroporation, letting
ROS and RNS molecules into the interior of the cells. These can then cause various deleterious effects
including DNA strand breaks and mitochondrial damage.

This book is a compilation of two special issues guest edited by Dr. Mounir Laroussi for the
journal *Plasma*. The two special issues are: (1) Low Temperature Plasma Jets: Physics, Diagnostics,
and Applications; (2) Plasma Medicine. This book is therefore organized into two parts. The first part
is a collection of six papers published in the special issues on plasma jets that discuss the design of
plasma jets such the Cooperation in Science and Technology (COST) plasma jets [46], the interaction
of plasma jets with various targets [47–49], characterizations of plasma jets, treatments of water by a
plasma jet [50], and the use of a plasma jet as a source of guided ionization waves to ignite a large
volume plasma in an electrodeless chamber [51]. The second part of the book is a collection of 8 papers
published in the special issue on plasma medicine. The first paper is an introductory review of the
field of plasma medicine [52]; some of the following papers discuss the applications of various plasma
jets for cancer treatment, including triple-negative breast cancer cells, ovarian cancer cells, and the
manner in which plasma can decrease the viability of malignant solid tumors cells [53–55]. One paper
presents a performance evaluation of three plasma sources/jets [56], while two other papers discuss
how plasma modulates the responsiveness of human macrophages and cellular glucose uptake [57,58].
Finally, the issue concludes with a review paper discussing how low temperature plasma offers a new
hope for cancer treatment [59].

Conflicts of Interest: The author declares no conflict of interest.

References

1. Von Siemens, W. Ueber die elektrostatische Induction und die Verzögerung des Stroms in Flaschendrähten. *Ann. Phys. Chem.* **1857**, *12*, 66. [CrossRef]
2. Kogelschatz, U. Silent discharges for the generation of ultraviolet and vacuum ultraviolet excimer radiation. *Pure Appl. Chem.* **1990**, *62*, 1667. [CrossRef]
3. Kogelschatz, U.; Eliasson, B.; Egli, W. Dielectric barrier discharges: Principle and applications. *J. Phys.* **1997**, *C4*, 47. [CrossRef]
4. Kogelschatz, U. Filamentary, patterned, and diffuse barrier discharges. *IEEE Trans. Plasma Sci.* **2002**, *30*, 1400. [CrossRef]
5. Laroussi, M. Sterilization of contaminated matter with an atmospheric pressure plasma. *IEEE Trans. Plasma Sci.* **1996**, *24*, 1188. [CrossRef]
6. Laroussi, M.; Lu, X. Room temperature atmospheric pressure plasma plume for biomedical applications. *Appl. Phys. Lett.* **2005**, *87*, 113902. [CrossRef]
7. Laroussi, M.; Akan, T. Arc-free atmospheric pressure cold plasma jets: A review. *Plasma Process. Polym.* **2007**, *4*, 777. [CrossRef]
8. Cheng, C.; Liye, Z.; Zhan, R. Surface modification of polymer fiber by the new atmospheric pressure cold plasma jet. *Surf. Coat. Technol.* **2006**, *200*, 6659. [CrossRef]
9. Chen, G.; Chen, S.; Zhou, M.; Feng, W.; Gu, W.; Yang, S. The preliminary discharging characterization of a novel APGD plume and its application in organic contaminant degradation. *Plasma Sources Sci. Technol.* **2006**, *15*, 603. [CrossRef]
10. Inomata, K.; Koinuma, H.; Oikawa, Y.; Shiraishi, T. Open air photoresist ashing by cold plasma torch: Catalytic effect of cathode material. *Appl. Phys. Lett.* **1995**, *66*, 2188. [CrossRef]
11. Brandenburg, R.; Ehlbeck, J.; Stieber, M.V.; von Woedtke, T.; Zeymer, J.; Schluter, O.; Weltmann, K.-D. Antimicrobial treatment of heat sensitive materials by means of atmospheric pressure rf-driven Plasma Jet. *Contrib. Plasma Phys.* **2007**, *47*, 72. [CrossRef]

12. Lu, X.; Reuter, S.; Laroussi, M.; Liu, D. *Non-Equilibrium Atmospheric Pressure Plasma Jets: Fundamentals, Diagnostics, and Medical Applications*; CRC Press: Boca Raton, FL, USA, 2019; ISBN 9781498743631.

13. Fridman, G.; Brooks, A.; Galasubramanian, M.; Fridman, A.; Gutsol, A.; Vasilets, V.; Ayan, H.; Friedman, G. Comparison of direct and indirect effects of non-thermal atmospheric-pressure plasma on bacteria. *Plasma Process. Polym.* **2007**, *4*, 370. [CrossRef]

14. Shashurin, A.; Keidar, M.; Bronnikov, S.; Jurjus, R.A.; Stepp, M.A. Living tissue under treatment of cold plasma atmospheric jet. *Appl. Phys. Lett.* **2008**, *93*, 181501. [CrossRef]

15. Yan, X.; Zou, F.; Zhao, S.; Lu, X.; He, G.; Xiong, Z.; Xiong, Q.; Zhao, Q.; Deng, P.; Huang, J.; et al. On the Mechanism of Plasma Inducing Cell Apoptosis. *IEEE Trans Plasma Sci.* **2010**, *38*, 9. [CrossRef]

16. Xiong, Z.; Cao, Y.; Lu, X.; Du, T. Plasmas in tooth root canal. *IEEE Trans Plasma Sci.* **2011**, *39*, 2968. [CrossRef]

17. Zimmermann, J.L.; Shimizu, T.; Boxhammer, V.; Morfill G., E. Disinfection through different textiles using low-temperature atmospheric pressure plasma. *Plasma Process. Polym.* **2012**, *9*, 792. [CrossRef]

18. Babaeva, N.; Kushner, M.J. Reactive fluxes delivered by dielectric barrier discharge filaments to slightly wounded skin. *J. Phys. D Appl. Phys.* **2013**, *46*, 025401. [CrossRef]

19. Weltmann, K.D.; Kindel, E.; Brandenburg, R.; Meyer, C.; Bussiahn, R.; Wilke, C.; Von Woedtke, T. Atmospheric Pressure Plasma Jet for Medical Therapy: Plasma Parameters and Risk Estimation. *Contrib. Plasma to Plasma Phys.* **2009**, *49*, 631. [CrossRef]

20. Ehlbeck, J.; Schnabel, U.; Polak, M.; Winter, J.; Von Woedtke, T.; Brandenburg, R.; Von dem Hagen, T.; Weltmann, K.D. Low temperature atmospheric pressure plasma sources for microbial decontamination. *J. Phys. D: Appl. Phys.* **2011**, *44*, 013002. [CrossRef]

21. Utsumi, F.; Kjiyama, H.; Nakamura, K.; Tanaka, H.; Mizuno, M.; Ishikawa, K.; Kondo, H.; Kano, H.; Hori, M.; Kikkawa, F. Effect of Indirect Nonequilibrium Atmospheric Pressure Plasma on Anti-Proliferative Activity against Chronic Chemo-Resistant Ovarian Cancer Cells In Vitro and In Vivo. *PLoS ONE* **2013**, *8*, e81576. [CrossRef]

22. Tanaka, H.; Mizuno, M.; Ishikawa, K.; Takeda, K.; Nakamura, K.; Utsumi, F.; Kajiyama, H.; Kano, H.; Okazaki, Y.; Toyokuni, S.; et al. Plasma Medical Science for Cancer Therapy: Toward Cancer Therapy Using Nonthermal Atmospheric Pressure Plasma. *IEEE Trans. Plasma Sci.* **2014**, *42*, 3760. [CrossRef]

23. Vandamme, M.; Robert, E.; Pesnele, S.; Barbosa, E.; Dozias, S.; Sobilo, J.; Lerondel, S.; Le Pape, A.; Pouvesle, J.-M. Antitumor Effects of Plasma Treatment on U87 Glioma Xenografts: Preliminary Results. *Plasma Process. Polym.* **2010**, *7*, 264. [CrossRef]

24. Fridman, G.; Shereshevsky, A.; Jost, M.M.; Brooks, A.D.; Fridman, A.; Gutsol, A.; Vasilets, V.; Friedman, G. Floating electrode dielectric barrier discharge plasma in air promoting apoptotic behavior in melanoma skin cancer cell lines. *Plasma Chem. Plasma Process.* **2007**, *27*, 163. [CrossRef]

25. Volotskova, O.; Hawley, T.S.; Stepp, M.A.; Keidar, M. Targeting the cancer cell cycle by cold atmospheric plasma. *Sci Rep-Uk* **2012**, *2*. [CrossRef] [PubMed]

26. Brandenburg, R. Dielectric barrier discharges: Progress on plasma sources and on the understanding of regimes and single filaments. *Plasma Sources Sci. Technol.* **2017**, *26*, 053001. [CrossRef]

27. Weltmann, K.-D.; Kindel, E.; von Woedtke, T.; Hähnel, M.; Stieber, M.; Brandenburg, R. Atmospheric-pressure plasma sources: Prospective tools for plasma medicine. *Pure Appl. Chem.* **2010**, *82*, 1223. [CrossRef]

28. Teschke, M.; Kedzierski, J.; Finantu-Dinu, E.G.; Korzec, D.; Engemann, J. High-speed photographs of a dielectric barrier atmospheric pressure Plasma Jet. *IEEE Trans. Plasma Sci.* **2005**, *33*, 310. [CrossRef]

29. Lu, X.; Laroussi, M. Dynamics of an atmospheric pressure plasma plume generated by submicrosecond voltage pulses. *J. Appl. Phys.* **2006**, *100*, 063302. [CrossRef]

30. Sands, B.L.; Ganguly, B.N.; Tachibana, K.A. Streamer-like atmospheric pressure plasma jet. *Appl. Phys. Lett.* **2008**, *92*, 151503. [CrossRef]

31. Mericam-Bourdet, N.; Laroussi, M.; Begum, A.; Karakas, E. Experimental investigations of plasma bullets. *J. Phys. D Appl. Phys.* **2009**, *42*, 055207. [CrossRef]

32. Naidis, G.V. Modeling of plasma bullet propagation along a helium jet in ambient air. *J. Phys. D Appl. Phys.* **2011**, *44*, 215203. [CrossRef]

33. Yousfi, M.; Eichwald, O.; Merbahi, N.; Jomma, N. Analysis of ionization wave dynamics in low-temperature plasma jets from fluid modeling supported by experimental investigations. *Plasma Sources Sci. Technol.* **2012**, *21*, 045003. [CrossRef]

34. Boeuf, J.-P.; Yang, L.; Pitchford, L. Dynamics of guided streamer (plasma bullet) in a helium jet in air at atmospheric pressure. *J. Phys. D Appl. Phys.* **2013**, *46*, 015201. [CrossRef]

35. Lu, X.; Naidis, G.; Laroussi, M.; Ostrikov, K. Guided ionization waves: Theory and experiments. *Phys. Rep.* **2014**, *540*, 123. [CrossRef]

36. Begum, A.; Laroussi, M.; Pervez, M.R. Atmospheric Pressure helium/air plasma Jet: Breakdown processes and propagation phenomenon. *AIP Adv.* **2013**, *3*, 062117. [CrossRef]

37. Stretenovic, G.B.; Krstic, I.B.; Kovacevic, V.V.; Obradovic, A.M.; Kuraica, M.M. Spatio-temporally resolved electric field measurements in helium plasma jet. *J. Phys. D Appl. Phys.* **2014**, *47*, 102001. [CrossRef]

38. Sobota, A.; Guaitella, O.; Garcia-Caurel, E. Experimentally obtained values of electric field of an atmospheric pressure plasma jet impinging on a dielectric surface. *J. Phys. D Appl. Phys.* **2013**, *46*, 372001. [CrossRef]

39. Graves, D. The emerging role of reactive oxygen and nitrogen species in redox biology and some implications for plasma applications to medicine and biology. *J. Phys. D: Appl. Phys.* **2012**, *45*, 263001. [CrossRef]

40. Laroussi, M. Low temperature plasma jet for biomedical applications: A review. *IEEE Trans. Plasma Sci.* **2015**, *43*, 703. [CrossRef]

41. Lu, X.; Naidis, G.V.; Laroussi, M.; Reuter, S.; Graves, D.B.; Ostrikov, K. Reactive species in non-equilibrium atmospheric pressure plasma: Generation, transport, and biological effects. *Phys. Rep.* **2016**, *630*, 1. [CrossRef]

42. Lu, X.; Keidar, M.; Laroussi, M.; Choi, E.; Szili, E.J.; Ostrikov, K. Transcutaneous plasma stress: From soft-matter models to living tissues. *Mater. Sci. Eng. R Rep.* **2019**, *138*, 36. [CrossRef]

43. Keidar, M.; Walk, R.; Shashurin, A.; Srinivasan, P.; Sandler, A.; Dasgupta, S.; Ravi, R.; Guerrero-Preston, R.; Trink, B. Cold plasma selectivity and the possibility of a paradigm shift in cancer therapy. *Br. J. Cancer* **2011**, *105*, 1295. [CrossRef] [PubMed]

44. Schlegel, J.; Koritzer, J.; Boxhammer, V. Plasma in cancer treatment. *Clin. Plasma Med.* **2013**, *1*, 2. [CrossRef]

45. Laroussi, M. Effects of PAM on select normal and cancerous epithelial cells. *Plasma Res. Express* **2019**, *1*, 025010. [CrossRef]

46. Gorbanev, Y.; Golda, J.; Gathen, V.; Bogaerts, A. Applications of the COST Plasma Jet: More than a Reference Standard. *Plasma* **2019**, *2*, 316. [CrossRef]

47. Teschner, T.; Bansemer, R.; Weltmann, K.; Gerling, T. Investigation of Power Transmission of a Helium Plasma Jet to Different Dielectric Targets Considering Operating Modes. *Plasma* **2019**, *2*, 348. [CrossRef]

48. Bolouki, N.; Hsieh, J.; Li, C.; Yang, Y. Emission Spectroscopic Characterization of a Helium Atmospheric Pressure Plasma Jet with Various Mixtures of Argon Gas in the Presence and the Absence of De-Ionized Water as a Target. *Plasma* **2019**, *2*, 283. [CrossRef]

49. Simoncelli, E.; Stancampiano, A.; Boselli, M.; Gherardi, M.; Colombo, V. Experimental Investigation on the Influence of Target Physical Properties on an Impinging Plasma Jet. *Plasma* **2019**, *2*, 369. [CrossRef]

50. Groele, J.; Foster, J. Hydrogen Peroxide Interference in Chemical Oxygen Demand Assessments of Plasma Treated Waters. *Plasma* **2019**, *2*, 294. [CrossRef]

51. Laroussi, M. Ignition of A Plasma Discharge Inside An Electrodeless Chamber: Methods and Characteristics. *Plasma* **2019**, *2*, 380. [CrossRef]

52. Laroussi, M. Plasma Medicine: A Brief Introduction. *Plasma* **2018**, *1*, 47. [CrossRef]

53. Cheng, X.; Rowe, W.; Ly, L.; Shashurin, A.; Zhuang, T.; Wigh, S.; Basadonna, G.; Trink, B.; Keidar, M.; Canady, J. Treatment of Triple-Negative Breast Cancer Cells with the Canady Cold Plasma Conversion System: Preliminary Results. *Plasma* **2018**, *1*, 218. [CrossRef]

54. Bekeschus, S.; Wulf, C.; Freund, E.; Koensgen, D.; Mustea, A.; Weltmann, K.; Stope, M. Plasma Treatment of Ovarian Cancer Cells Mitigates Their Immuno-Modulatory Products Active on THP-1 Monocytes. *Plasma* **2018**, *1*, 201. [CrossRef]

55. Rowe, W.; Cheng, X.; Ly, L.; Zhuang, T.; Basadonna, G.; Trink, B.; Keidar, M.; Canady, J. The Canady Helios Cold Plasma Scalpel Significantly Decreases Viability in Malignant Solid Tumor Cells in a Dose-Dependent Manner. *Plasma* **2018**, *1*, 177. [CrossRef]

56. Ly, L.; Jones, S.; Shashurin, A.; Zhuang, T.; Rowe, W.; Cheng, X.; Wigh, S.; Naab, T.; Keidar, M.; Canady, J. A New Cold Plasma Jet: Performance Evaluation of Cold Plasma, Hybrid Plasma and Argon Plasma Coagulation. *Plasma* **2018**, *1*, 189. [CrossRef]

57. Crestale, L.; Laurita, R.; Liguori, A.; Stancampiano, A.; Talmon, M.; Bisag, A.; Gherardi, M.; Amoruso, A.; Colombo, V.; Fresu, L. Cold Atmospheric Pressure Plasma Treatment Modulates Human Monocytes/Macrophages Responsiveness. *Plasma* **2018**, *1*, 261. [CrossRef]

58. Razzokov, J.; Yusupov, M.; Bogaerts, A. Possible Mechanism of Glucose Uptake Enhanced by Cold Atmospheric Plasma: Atomic Scale Simulations. *Plasma* **2018**, *1*, 119. [CrossRef]
59. Tanaka, H.; Mizuno, M.; Ishikawa, K.; Toyokuni, S.; Kajiyama, H.; Kikkawa, F.; Hori, M. New Hopes for Plasma-Based Cancer Treatment. *Plasma* **2018**, *1*, 150. [CrossRef]

Mounir Laroussi

 *plasma*

Editorial

Special Issue on Low Temperature Plasma Jets

Mounir Laroussi

Electrical & Computer Engineering Department, Old Dominion University, Norfolk, VA 23529, USA; mlarouss@odu.edu; Tel.: 757-683-6369

Received: 12 July 2019; Accepted: 30 July 2019; Published: 31 July 2019

Low temperature plasma jets are unique plasma sources capable of delivering plasma outside of the confinement of electrodes and away from gas enclosures/chambers. With these jets plasma can be easily delivered to a target located at some distance from the plasma generation region [1]. Various power driving methods have been used to ignite and sustain low temperature plasma jets. These include direct current (DC), pulsed DC, Radio Frequency (RF), and microwave power [1]. In particular, low temperature plasma jets that are generated at atmospheric pressure are playing an ever increasing role in many plasma processing applications, including surface treatment and in biomedicine. This is because they provide interesting reactive chemistry that can be exploited in various processing applications. Reactive oxygen and nitrogen species (RONS), such as O, OH, O_2^-, $^1O_2$, H_2O_2, NO, NO_2, which are generated by these plasma jets, have been shown to play a central role in their interactions with solids surfaces, liquids, and soft matter (including cells and tissues).

The discovery that atmospheric pressure, low temperature plasma jets are in fact not continuous plasma plumes but fast propagating discrete small volumes of plasma (known as "plasma bullets") makes the physics of these jets particularly interesting [2,3]. This led to numerous experimental and modeling works which aimed at elucidating their mechanisms of ignition and propagation [4–8]. Today, it is well established that these jets are enabled by guided ionization waves where photoionization and the electric field at the head of the ionization front play important roles [9]. The magnitude of the electric field was measured by several investigators and was found to be in the 10–30 kV/cm range [10–13].

Low temperature plasma jets have been used in various applications. For example, in material processing, using various operating conditions and gases, they were found to increase the wettability of Polypropylene (PP) and Polyethylene terephthalate (PET) films [14], degrade aromatic rings of dies such methyl violet [15], etch silicon, Si (100), ash photoresist at a rate greater than 1.2 µm/min [16], deposit silicon dioxide, SiO_2, films on various substrates at deposition rates greater than 10 nm/s, etc. However, and by far, the most interesting and emerging applications of low temperature plasma jets are in biomedicine. In this field of research, intense investigations of their various biomedical applications surged ever since the first "bio-tolerant" plasma jets were reported in the mid-2000s [17,18]. Today these plasma jets are being extensively researched for medical applications ranging from wound healing, to dentistry, and to cancer therapy [19–21].

This special issue contains articles discussing the latest works which cover both fundamental studies and applications of low temperature plasma jets. The guest editor would like to thank the authors for their valuable contributions and the reviewers for their time and efforts.

References

1. Laroussi, M.; Akan, T. Arc-free Atmospheric Pressure Cold Plasma Jets: A Review. *Plasma Process. Polym.* **2007**, *4*, 777. [CrossRef]
2. Teschke, M.; Kedzierski, J.; Finantu-Dinu, E.G.; Korzec, D.; Engemann, J. High-Speed Photographs of a Dielectric Barrier Atmospheric Pressure Plasma Jet. *IEEE Trans. Plasma Sci.* **2005**, *33*, 310. [CrossRef]

3. Lu, X.; Laroussi, M. Dynamics of an Atmospheric Pressure Plasma Plume Generated by Submicrosecond Voltage Pulses. *J. Appl. Phys.* **2006**, *100*, 063302. [CrossRef]

4. Sands, B.L.; Ganguly, B.N.; Tachibana, K.A. Streamer-like Atmospheric Pressure Plasma Jet. *Appl. Phys. Lett.* **2008**, *92*, 151503. [CrossRef]

5. Mericam-Bourdet, N.; Laroussi, M.; Begum, A.; Karakas, E. Experimental Investigations of Plasma Bullets. *J. Phys. D Appl. Phys.* **2009**, *42*, 055207. [CrossRef]

6. Naidis, G.V. Modeling of Plasma Bullet Propagation along a Helium Jet in Ambient Air. *J. Phys. D Appl. Phys.* **2011**, *44*, 215203. [CrossRef]

7. Yousfi, M.; Eichwald, O.; Merbahi, N.; Jomma, N. Analysis of ionization wave dynamics in low-temperature plasma jets from fluid modeling supported by experimental investigations. *Plasma Sources Sci. Technol.* **2012**, *21*, 045003. [CrossRef]

8. Boeuf, J.-P.; Yang, L.; Pitchford, L. Dynamics of guided streamer (plasma bullet) in a helium jet in air at atmospheric pressure. *J. Phys. D Appl. Phys.* **2013**, *46*, 015201. [CrossRef]

9. Lu, X.; Naidis, G.; Laroussi, M.; Ostrikov, K. Guided Ionization Waves: Theory and Experiments. *Phys. Rep.* **2014**, *540*, 123. [CrossRef]

10. Begum, A.; Laroussi, M.; Pervez, M.R. Atmospheric Pressure helium/air plasma Jet: Breakdown Processes and Propagation Phenomenon. *AIP Adv.* **2013**, *3*, 062117. [CrossRef]

11. Stretenovic, G.B.; Krstic, I.B.; Kovacevic, V.V.; Obradovic, A.M.; Kuraica, M.M. Spatio-temporally resolved electric field measurements in helium plasma jet. *J. Phys. D Appl. Phys.* **2014**, *47*, 102001. [CrossRef]

12. Sobota, A.; Guaitella, O.; Garcia-Caurel, E. Experimentally obtained values of electric field of an atmospheric pressure plasma jet impinging on a dielectric surface. *J. Phys. D Appl. Phys.* **2013**, *46*, 372001. [CrossRef]

13. Darny, T.; Pouvesle, J.-M.; Puech, V.; Douat, C.; Dozias, S.; Robert, E. Analysis of conductive target influence in plasma jet experiments through helium metastable and electric field measurements. *Plasma Sources Sci. Technol.* **2017**, *26*, 045008. [CrossRef]

14. Cheng, C.; Liye, Z.; Zhan, R. Surface Modification of Polymer Fiber by the New Atmospheric Pressure Cold Plasma Jet. *Surface Coat. Technol.* **2006**, *200*, 6659. [CrossRef]

15. Chen, G.; Chen, S.; Zhou, M.; Feng, W.; Gu, W.; Yang, S. The Preliminary Discharging Characterization of a Novel APGD Plume and its Application in Organic Contaminant Degradation. *Plasma Sources Sci. Technol.* **2006**, *15*, 603. [CrossRef]

16. Inomata, K.; Koinuma, H.; Oikawa, Y.; Shiraishi, T. Open Air Photoresist Ashing by Cold Plasma Torch: Catalytic effect of Cathode material. *Appl. Phys. Lett.* **1995**, *66*, 2188. [CrossRef]

17. Laroussi, M.; Lu, X. Room Temperature Atmospheric Pressure Plasma Plume for Biomedical Applications. *Appl. Phys. Lett.* **2005**, *87*, 113902. [CrossRef]

18. Brandenburg, R.; Ehlbeck, J.; Stieber, M.V.; von Woedtke, T.; Zeymer, J.; Schluter, O.; Weltmann, K.-D. Antimicrobial Treatment of Heat Sensitive Materials by Means of Atmospheric Pressure rf-driven Plasma Jet. *Contrib. Plasma Phys.* **2007**, *47*, 72. [CrossRef]

19. Keidar, M.; Shashurin, A.; Volotskova, O.; Stepp, M.A.; Srinivasan, P.; Sandler, A.; Trink, B. Cold atmospheric plasma in cancer therapy. *Phys. Plasmas* **2013**, *20*, 057101. [CrossRef]

20. Gherardi, M.; Tonini, R.; Colombo, V. Plasma in Dentistry: Brief History and Current Status. *Trends Biotechnol.* **2017**, *36*, 583. [CrossRef] [PubMed]

21. Lu, X.; Reuter, S.; Laroussi, M.; Liu, D. *Non-Equilibrium Atmospheric Pressure Plasma Jets: Fundamentals, Diagnostics, and Medical Applications*; CRC Press: Boca Raton, FL, USA, 2019; ISBN 9781498743631.

Review

Applications of the COST Plasma Jet: More than a Reference Standard

Yury Gorbanev [1,*], Judith Golda [2,3], Volker Schulz-von der Gathen [3] and Annemie Bogaerts [1]

1   Research group PLASMANT, Department of Chemistry, University of Antwerp, Universiteitsplein 1, 2610 Wilrijk, Belgium
2   Institute for Experimental and Applied Physics, Kiel University, Leibnizstraße 19, 24118 Kiel, Germany
3   Chair for Experimental Physics II: Reactive plasmas, Ruhr-University Bochum, Universitätsstraße 150, 44801 Bochum, Germany
*   Correspondence: yury.gorbanev@uantwerpen.be; Tel.: +32-(0)-326-52-343

Received: 26 June 2019; Accepted: 10 July 2019; Published: 12 July 2019

Abstract: The rapid advances in the field of cold plasma research led to the development of many plasma jets for various purposes. The COST plasma jet was created to set a comparison standard between different groups in Europe and the world. Its physical and chemical properties are well studied, and diagnostics procedures are developed and benchmarked using this jet. In recent years, it has been used for various research purposes. Here, we present a brief overview of the reported applications of the COST plasma jet. Additionally, we discuss the chemistry of the plasma-liquid systems with this plasma jet, and the properties that make it an indispensable system for plasma research.

Keywords: COST microplasma jet; reference plasma jet; µ-APPJ; plasma chemistry; plasma applications; biomedical plasma; plasma-liquid interactions; plasma RONS; plasma-polymer interactions; nanomaterials

---

1. Introduction

Among the various types of plasma, non-thermal (or 'cold') atmospheric pressure plasma (CAP) is perhaps the most burgeoning field [1,2]. CAPs find their applications in biomedical, chemical, environmental/energy and industrial research [1,3–5]. As a result of the extensive range of applications, a plethora of plasma devices have been developed and reported in literature.

In turn, among the various types of plasma setups, atmospheric pressure plasma jets (APPJs) are some of the most widely used devices due to their unique properties [6,7]. Operated at ambient temperature and pressure, they enable direct treatment of temperature-sensitive substrates, including biological targets (cells, tissues, agricultural materials, etc.). APPJs in many cases have a minimised electrical impact during plasma treatment, while at the same time facilitating targeted delivery of the biologically active reactive oxygen and nitrogen species (RONS) due to the flow of gas. These RONS comprise long-lived ones (molecules and ions) and short-lived ones (radicals and atoms), and define the potential of CAPs in (among others) biomedical applications [8,9].

APPJs are operated with a flow of feed gas between the electrodes. This flow of gas is responsible for the term 'jet'. The feed gas is usually an inert gas (e.g., Ar or He, pure or with added molecular admixtures [10,11]), although in certain cases nitrogen or air is used [12]. The RONS are created either inside the jet, or when the effluent of the jet interacts with the ambient atmosphere [13]. Furthermore, APPJs can be discriminated based on the parameters of the discharge (pulsed or continuous sinusoidal), frequency (e.g., kHz, MHz), and configuration of the electrodes, etc. The electrode configuration allows distinguishing between the two types of APPJs: Parallel field and cross field APPJs. In the former, the applied electric field is parallel to the gas flow, and in the latter, it is perpendicular [3,7,14].

Thus, there are numerous differences in properties and effects of CAP applications even within the various APPJs. Despite being necessary for the field to progress, this variety of APPJs creates difficulties in comparison between results, and in deconvolution of plasma effects.

To address these issues, within the European Cooperation for Science and Technology (COST) action MP1101 'Biomedical Applications of Atmospheric Pressure Plasmas' [15] the COST reference Microplasma Jet was developed from its predecessor, the μ-APPJ.

In short, the COST jet comprises two stainless steel electrodes of 30 mm length and 1 mm width. Extensions allow the connection with the power supply, as shown in Figure 1. The distance between the electrodes is 1 mm. The electrodes are sealed between quartz panes, thus forming a discharge volume of 30 mm^3. Feed gas, typically a mixture of helium and a molecular admixture in the percent range, is introduced into this electrode stack through the gas connector made from ceramics. Gas flows in the range of 0.25 to several standard litres per min (slm) yield stable operation. Molecular admixtures can range up to a few vol% depending on the type of admixture. For the standard gas flow of 1 slm, an effluent velocity of about 15 m/s is obtained. The length of the electrodes ensures that a plasma chemical equilibrium is established in the discharge region before the feed gas leaves the plasma jet [16,17]. The evolution of the equilibrium throughout the complete plasma channel can be investigated and surveyed for the COST jet due to the direct optical access through the quartz panes.

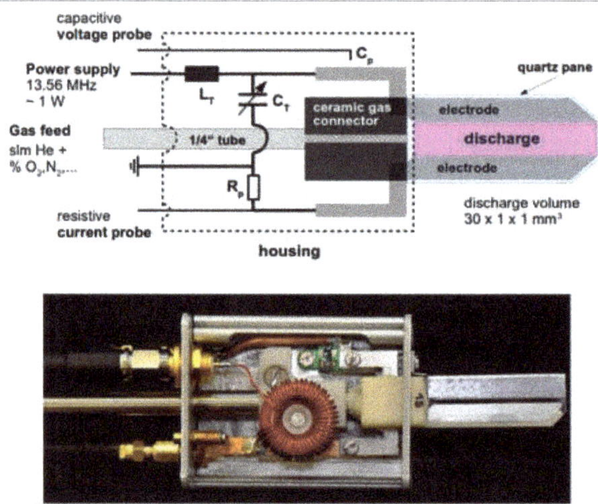

Figure 1. The schematics (top) and a photograph (bottom) of the European Cooperation for Science and Technology (COST) plasma jet.

We note that these parameters and design features are based on results from the thoroughly investigated predecessor—the μ-APPJ. The configuration of that is very similar, allowing to assume comparable results. The COST jet is operated with a capacitively coupled RF frequency of 13.56 MHz at ca. 1 W and a voltage of ca. 200–250 V$_{RMS}$.

To ensure and control proper operation, two probes are integrated into the COST-jet design directly connected to the electrodes. For current measurements, a precision resistor (R$_P$) is used. Voltage measurements are realised by a pin probe (C$_P$) at the powered electrode. The combined measurement of the current, voltage and phase allows the measurement of the power input into the electrode stack. The latter together with the inductance L$_T$ and the tunable capacitance C$_T$ form a part of a resonance circuit that allows the use of a low power, low voltage power supply providing sinusoidal waveforms. Other supplies can be fitted that allow for other waveforms and frequencies or pulsed operation of the jet, although they are not included in the original COST jet definition. All components are installed inside a grounded housing that shields the whole assembly against external stray capacitances.

Further details of the physical properties, as well as the electrical and gas phase plasma diagnostics of the COST jet, and procedure protocols for reproducible operation are comprehensively described in an earlier work [18]. A set of modified COST jet configurations was developed and characterized that allows the separation of the particle and photons to study their isolated or combined effects [19]. Moreover, accurate computational models have been developed to predict the chemical kinetics in the gas phase [20–22].

Overall, the possibility to control, tune and reproduce the gas phase plasma properties of the COST jet and modified jets make them highly useful plasma setups.

In this review, we summarise the main advances in the applications of the COST plasma jet. Specifically, we focus on the insights into plasma chemistry gained by using the COST jet. Here we report the use of the COST jet together with its predecessor μ-APPJ, which have virtually the same properties, as mentioned above. Henceforth, we refer to both as 'the COST jet' to maintain clarity.

2. Applications of the COST Plasma Jet

2.1. Interaction with Organic Polymers

The applicability of an APPJ to specific purposes is largely defined by the induced chemistry. The aforementioned tunability of the COST jet (and COST jet-based modified plasma setups) allows controlled generation of RONS in the gas phase [19,23]. Specifically, when operated with a feed gas of He with O_2 admixtures, the main 'output' reactive oxygen species (ROS) are atomic oxygen O, singlet oxygen $^1O_2$, and ozone O_3 [24,25]. These ROS are of high importance in the field of material research. For example, highly chemically reactive atomic and radical species generated by plasma [8,13] can interact with polymer surfaces in various ways.

2.1.1. Photoresist Removal

In the micro-nano-electronics industry, the production of lithographic patterns proceeds through several steps, required to generate advanced nanoscale structures. The removal of photoresistant polymers (or plasma etching) between the processing steps is used. Commonly, this is done by using low-pressure plasmas, the operation of which is costly and can damage the substrate surface [26,27].

An alternative to this method was suggested by West et al. [27]. Since in the COST jet the applied field is perpendicular to the gas flow, the charged species, including high energy electrons, are largely confined within the jet. This eliminates the danger of sheath formation when the effluent of the jet interacts with targets.

The COST jet was operated at 13.56 and 40.68 MHz, with a feed gas of He with O_2 admixtures. It was used to remove novolak polymers from a silicon wafer. It was found that under the investigated conditions, an admixture of 0.5% O_2 resulted in the highest rate of photoresist etching. The authors reported that 0.5% O_2 in He yielded the highest amount of atomic O, which also corresponded to the highest rate of etching. O was suggested to be the main species responsible for the polymer removal. The highest etching rate achieved by West et al. was 10 μm/min, with a total gas flow rate of 7 L/min. This etching rate is comparable to those required by the semiconductor industry, while being substantially more benign for the underlying surface.

Very recently, Hefny et al. showed that when the COST jet $He+O_2$ plasma interacts with surfaces, such as e.g., amorphous carbon films, the probability of the surface loss of O atoms is low (below 1%). This results in high densities of O atoms not only in the area of the immediate contact of the plasma effluent with the surface, but also radially on the surrounding thin film, making the COST jet a useful tool in studying the surface processes occurring during CAP etching [28].

2.1.2. Studying the Main Agents Used in Polymer Surface Modification

Polymers are widely used in the packaging industry, for textile, and in medical applications, often in combination with plasma technology [29]. The low wettability of polymers can be a problem that needs resolving. The increased hydrophilicity of the polymer surface is achieved by using industrial plasmas, but the chemistry which yields the desired properties of polymers is not well known [30].

Shaw et al. investigated the chemical species and mechanisms leading to the induction of wettability in polypropylene films [31]. Once again, here the COST jet was used as a system without an active effluent: the 'output' of the jet was comprised of highly chemically reactive atoms and radicals, but contained no high energy species such as plasma electrons.

The authors assessed the wettability from the contact angle between water droplets and the polymer surface as a function of various plasma parameters. Similar to the etching study, the laser induced fluorescence analysis showed that most O atoms were produced in the studied system when He was used with ca. 0.5% O_2. This O_2 admixture also corresponded to the highest wettability achieved. The surface analysis revealed the formation of C=O moieties corresponding to both ketones and carboxylic acids.

This demonstrates that the COST jet-induced chemistry (e.g., O-based) can be used both for complete etching of target polymers, and for mild oxidative transformation of the surface groups leading to increased wettability of polymers, depending on the specific parameters of plasma operation.

2.2. Preparation of Silicon-Based Films

Silicon dioxide is a common material for the production of thin films, which find their use in anti-corrosive or scratch-resistant coatings [32]. One of the most attractive ways to deposit these films is with CAPs, which allow reducing production costs by avoiding vacuum plasma technologies currently used [33].

Reuter et al. used the COST jet and its modified versions to study the plasma-driven deposition of silicon thin films with a specific chemical composition [32]. The plasma jet was operated with a feed gas of He with hexamethyldisiloxane (HMDSO). The authors investigated the effect of ROS present in the deposition process. O_2 either was added in a closed reactor with another jet (He+O_2, generating ROS), or diffused into the effluent from the ambient air. By studying the resulting film properties with FTIR, it was found that the carbon content of the deposited films depended largely on the presence of O_2: A He+HDMSO plasma jet yielded high C content films, while He+HDMSO with He+O_2 or with O_2 from the ambient air gave predominantly a Si-O-Si bond formation. The authors suggested that: (i) The He+HDMSO plasma jet produced Si-C rich thin films; (ii) they further interacted with atomic O or other ROS, leading to the loss of carbon. This was further confirmed by the ellipsometry and XPS analyses of the thin film produced by adding both HDMSO and O_2 into the He feed gas [34].

Further, Rügner et al. proposed reaction pathways leading to thin films with high Si-O and low Si-C content [35]. The addition of O_2 yielded ROS which lead to additional Si-O bonds in the gas phase plasma. This generates O-centred radical structures with a high affinity to surface deposition. Furthermore, the remaining methyl substituents on Si are removed via oxidation to CO_2, in turn leading to crosslinking and creating Si-O-Si structures (Scheme 1).

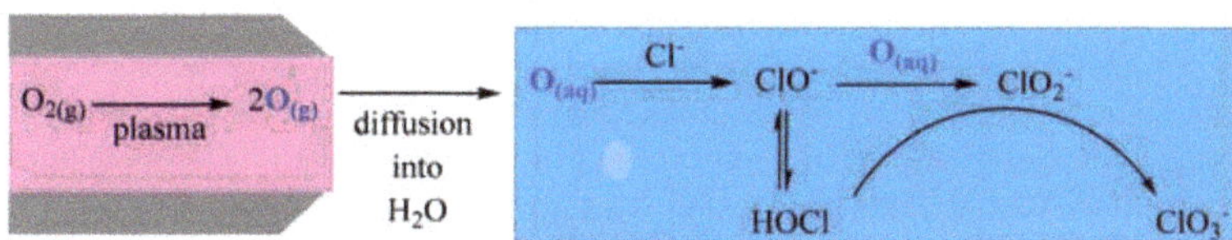

Scheme 1. The use of the COST jet to study the reactions leading to SiO_2-based thin films deposited by cold atmospheric pressure plasma (CAP).

Thus, the plasma generated with the COST jet proved useful in studying many fundamental properties of SiO_2 film deposition and the effect of other plasma RONS.

2.3. COST Jet for Inorganic and Organic Chemistry

One of the unique properties of the COST jet is its ability to produce ROS in the gas phase with high selectivity, as we discussed above. Additionally, and perhaps more importantly, it was also shown by studying the oxidative transformations of phenolic compounds in water that the COST jet is a very effective instrument to induce O atoms in liquid solutions [36,37]. In general, CAPs are unique systems for the delivery of short-lived species (atoms and radicals) [38], very sought-after in chemical applications, as we demonstrate below.

2.3.1. Studying the Reaction between Atomic O and Cl^- in the Liquid Phase

Due to the difficulties in selective induction of atomic oxygen into liquids, its reactions in aqueous solutions are largely unknown. One of such reactions is the oxidation of Cl^-, which was recently proposed by Kondeti et al. [39].

This reaction was studied using a COST jet operated with the $He+O_2$ feed gas. The produced plasma interacted with solutions containing Cl^- anions. We studied the production of various RONS: (i) Experimentally in aqueous solutions by measuring their concentrations with UV-Vis and EPR analyses, and (ii) computationally in the gas phase using 0D chemical kinetics modelling (enabled by the well-studied kinetic models of the gas phase plasma of the COST jet). The results showed nearly identical trends of atomic O densities in the gas phase and the ClO^- formed in the liquid phase. Importantly, the concentrations of other RONS which could potentially oxidise Cl^- (H_2O_2, O_3, $^1O_2$, •OH, •OOH) did not follow similar trends. This allowed us to suggest that ClO^- is formed via the direct oxidation of Cl^- by O atoms [25]. This was also confirmed by Jirásek et al., who used the COST jet with $He+O_2$ to treat chloride-rich solutions at high pH. The authors also observed the products of further oxidation: ClO_2^- and ClO_3^- [40] (Scheme 2). The reaction rate coefficient of O with Cl^- was estimated to be relatively 2–3 orders of magnitude lower than of O with 2,2,6,6-tetramethylpiperidine [25] and of O with phenol [40]. More accurate evaluation requires additional kinetic investigations.

Scheme 2. Use of the COST jet as an efficient source of oxygen (O) atoms to study reaction pathways in the oxidation of Cl^- in aqueous media.

These were the first studies of this previously undescribed chemical reaction, very important for fundamental chemistry research. They also have large implications on the in-liquid chemistry of CAP-treated solutions, especially in biomedical research.

2.3.2. Epoxidation of Trans-Stilbene with Atomic O

Epoxidation of olefins is an important chemical reaction yielding organic epoxides, key building blocks in organic synthesis. Epoxidations are typically performed with peroxy acids or other sacrificial oxygen donors, and/or catalysed by complex catalysts which are often not easily recyclable [41]. This creates separation, cost-effectiveness and environment-related issues by producing chemical waste.

In the context of benign, nearly zero-waste chemical agents, CAP-generated RONS present an attractive alternative to conventional chemical routes for organic synthesis [42].

Iza et al. studied the possibilities of epoxidation of *trans*-stilbene by various reactive species [43]. The authors utilised the COST jet operated with He+O_2 and He+CO_2. In both cases, the ROS that yielded the desired epoxide was the atomic O. This was shown by using an O_3 generator and $^1O_2$-producing photochemical system, neither of which produced large amounts of the epoxidation product. In the case of the He+CO_2 APPJ, the by-product of the reaction was CO. Direct plasma-driven CO_2 conversion to CO, although attracting much attention, is hindered by the reverse reaction of CO with O atoms [5]. *Trans*-stilbene in this case acted as a quencher for O, minimising this effect (Scheme 3). Exposing a solution of *trans*-stilbene in acetonitrile to these two effluents of the COST jet, the highest epoxide yield was ca. 60%, proving the feasibility of such waste- and catalyst-free epoxidation process.

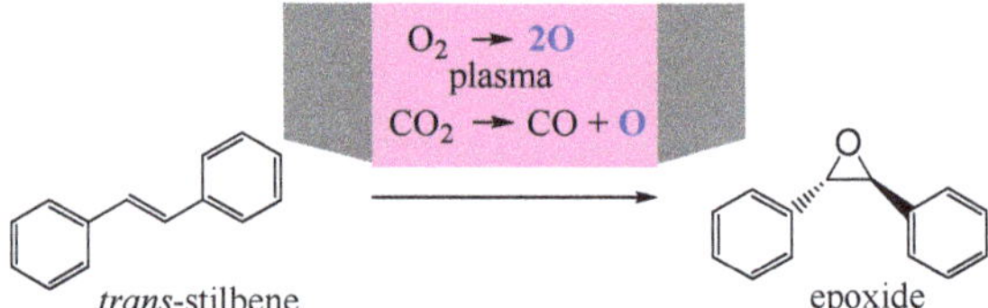

Scheme 3. Use of the COST jet as a waste-free benign chemical system for the epoxidation of *trans*-stilbene, providing high yield and conversion values.

2.4. Biomedical Research

Cold plasma medicine is one of the most extensive fields of CAP research. It comprises many applications, from cosmetic dentistry and biomedical materials production [1,9,29] to bacterial deactivation [39,44] and anti-cancer therapy [38,45].

Despite the significant progress in the field of biomedical CAP, the underlying mechanisms are far from understood. It is generally assumed that the majority of the observed biomedical effects is due to the extremely high chemical reactivity of RONS. However, the main RONS (or even their nature: long- or short-lived) responsible for the medicinal effects of CAP are mostly not known [38,46].

The deconvolution of the combined RONS effects becomes possible by using a CAP system with tunable production of reactive species, such as the COST jet. However, in biomedical milieu, RONS from the gas phase interact with liquid water present in every biological system. Therefore, studying the mechanisms of this interaction is required to evaluate the said tunability [10].

In a combined experimental and computational study, the sources of RONS induced by the COST jet in an exposed liquid were identified. This required differentiating between the water from the exposed liquid, water in the feed gas, and water in the effluent of the jet. It was done by isolating all components of the jet from the ambient air and using isotopically labelled water molecules [47]. As a result, it was found that even in the interaction with liquids, RONS produced by the COST jet are formed almost exclusively inside the plasma jet, from the components of the feed gas. In the effluent, these non-charged species further interact with each other and the components of air. Since the high energy species are absent in the COST jet effluent, virtually no new reactive species are formed from the air and H_2O vapour [48].

Therefore, even the small admixtures to the feed gas of the COST jet play a more important role than the composition of the ambient air (e.g., humidity level). This once again emphasises the 'standard' nature of the COST jet, which enables its usage in different environments comparably and controllably.

2.4.1. Studying Bactericidal Effects of CAPs

Treatments of chronic wounds with CAPs result in expedited healing, partly due to the efficient bacterial deactivation [9,44]. CAPs consist of different groups of physical and chemical components: (UV) photons, electric fields, charged and neutral particles [1,13]. Although numerous reports are available in literature on bacterial deactivation by exposure to cold plasma, the effect of these groups is not clearly understood.

Lackmann et al. used the COST jet to decouple the effects of various plasma-generated components on bacteria. The authors used modifications of the COST jet, which enabled the flux of (i) only photons; (ii) only RONS; (iii) combined fluxes of photons and RONS, while the gas flow onto the target remained the same under all experimental conditions. The most pronounced anti-bacterial effects were observed when the effluent with the combined flux was used: DNA damage, chemical modifications to protein and cell envelope were all the highest in this case. This indicated that the anti-bacterial effects of plasma can extend beyond the generation and delivery of RONS, and include the synergistic effects of UV photons and RONS combined [49].

Furthermore, the generalised effects of plasma RONS on proteins were studied using cysteine as a model molecule. Using a He-only fed COST jet, almost no changes in the cysteine structure were detected, whereas adding O_2 admixtures to He revealed multiple oxidation sites [50], expectedly due to the generation of highly oxidising O atoms. Here, the COST jet was used as a plasma device without an active effluent, as opposed to plasmas with high densities of high-energy particles in the effluent [10].

2.4.2. Identifying Optimal Parameters for Anti-CANCER treatments

CAP cancer therapy is one of the most promising anti-cancer modalities [9,45]. When combined with other treatments and surgical methods, the chance of the recurrence of tumour formation can be substantially reduced.

The number of studies on the CAP treatment of cancer cells has dramatically increased in recent years. Yet, similar to the anti-bacterial properties of cold plasma, the exact plasma components and the mechanisms responsible for the plasma effects are not explicitly known.

Vermeylen et al. used the COST jet operated with the He+O_2 feed gas to study the effects of various parameters on the treatment of different melanoma and glioblastoma cell lines [51]. While the viability of all cell types was reduced as a result of plasma treatment, some appeared to be more susceptible to it than the others. This study tackled one of the most highlighted problems that plasma medicine research is trying to solve: Determining the potential selectivity of CAP treatment on cancer cells compared to benign cells [52]. The COST jet was used under conditions producing high densities of O atoms.

Bekeschus et al. showed that the O atoms were one of the main RONS leading to the apoptotic death of leukemia cells [53]. The authors used the COST jet operated with the He with the H_2O and/or O_2 admixtures. The authors identified liquid phase reactions leading to ClO^- formation (see above) and other RONS leading to cell death. At the same time, H_2O_2, often considered to be the main biomedical effecter among the plasma-induced RONS [53,54], did not play an important role in the plasma-elicited effects.

The COST jet was also used in the comparative use of plasma-treated media as opposed to the direct plasma treatment [51]. This addresses one of the main reasons for the scepticism towards plasma medicine. Direct plasma treatment provides targeted delivery of highly short-lived RONS during treatment, while plasma-treated media generates persistent and semi-persistent molecular and ionic compounds, including e.g., H_2O_2, nitrous and peroxynitrous acid [13,55]. These compounds are commercially available, and their use technically does not require the employment of cold plasma.

A study was conducted determining the main RONS effecters on 3D tumour models (spheroids), which represent real in vivo tumours in a more realistic manner [56]. The spheroids consisted of glioblastoma cancer cells, and were subjected to treatments by the COST jet fed with the He feed gas with H_2O vapour admixtures. The main RONS induced by this APPJ in liquid were the H_2O_2,

NO_2^-, and $\bullet OH$ radicals. The effects of the direct plasma treatment and the CAP-treated media were compared. Despite the substantial short-term effects from the treated media (which was practically identical to the solutions prepared from the commercial chemicals such as H_2O_2), the long-term effects were only observed in the case of direct CAP treatment (Figure 2). Therefore, the short-lived radical species were proposed to be the key RONS for efficient tumour treatment. This indeed implies the need for plasma usage in cancer therapy, where plasma acts as a unique physicochemical system for the generation and delivery of short-lived RONS [38,56].

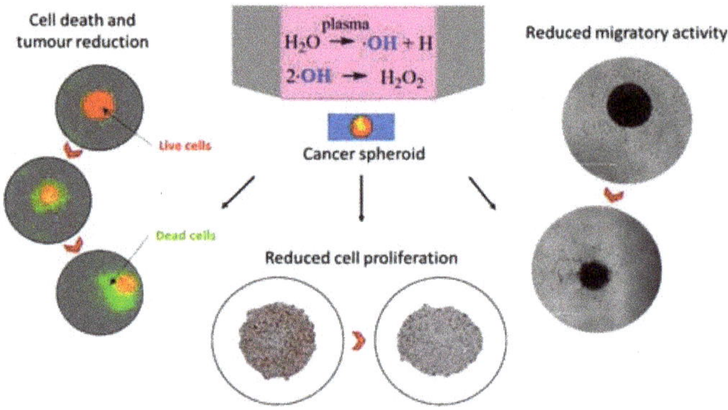

Figure 2. Strong long-term effects on glioblastoma spheroids were achieved only during direct CAP treatments due to the presence of short-lived species, such as $\bullet OH$ radicals, while the plasma-generated H_2O_2 resulted only in short-term effects.

3. Conclusions and Outlook

Cold plasma research is a vividly growing field. Different plasma devices are developed and optimised to achieve desired effects in specific applications. Among them is the COST plasma jet: A highly tunable, computationally approachable, and easy-to-use device. In this review, we showed its applications in several studies, such as materials research (polymer etching, thin film preparation) and biomedicine (anti-bacterial and anti-cancer treatments).

The ability of the COST jet to selectively produce RONS under specific conditions is what extends its applicability beyond being a standard reference plasma jet, and makes it an extremely useful tool for CAP-related investigations. For example, $O/^1O_2/O_3$ can be produced when O_2 is used as the feed gas admixture, and even then, the ratio between e.g., O and O_3 is easily tailored by further altering plasma exposure conditions; H_2O admixtures produce nearly exclusively H_2O_2, $\bullet OH$ and $\bullet OOH$, etc.

This brings us to the main fundamental difference between a cross-field plasma jet, such as the COST jet, and other types of CAP, such as dielectric barrier discharges (DBDs) or parallel-field APPJs. In the COST jet, the charged species, including high energy electrons, are contained within the plasma discharge region between the electrodes. The plasma electrons are responsible for most of the reactive non-equilibrium chemistry, which yields RONS upon interaction with the ambient air in the plasma jet effluent. Therefore, the effect of ambient conditions is dramatically reduced when using the COST jet compared to e.g., a DBD plasma or a parallel-field plasma jet. This also means that the COST jet is capable of producing RONS rather selectively, as opposed to e.g., the active effluent of parallel-field APPJs, where different RONS are produced from N_2, O_2 and H_2O at the same time, making the total number of different RONS high. In other words, the tunable and controllable RONS-producing chemistry occurring inside the COST jet can be used to study the effects of specific reactive species in chemical and biological systems, as we show in this review. Furthermore, the absence of active effluent enables the COST jet use in studies which aim to avoid damaging the target (e.g., plasma etching/ashing, biomedical work). Some of the most important studies are summarised in Figure 3.

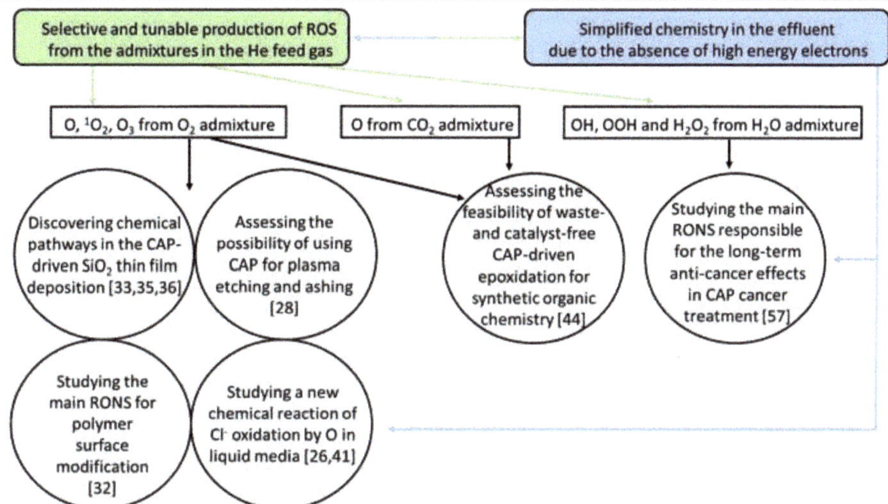

Figure 3. Some of the applied and fundamental research works with the COST plasma jet.

Several of the example works discussed here show the feasibility of using a CAP device, and specifically the COST jet, for industrial applications, e.g., in plasma etching, thin film deposition, and epoxidation in synthetic chemistry. However, it must be acknowledged that the direct use of the COST jet in an industrial or a clinical setting is limited due to several factors. For example, the COST jet requires the use of He as a (costly) feed gas. Nonetheless, we stress that it is indispensable in performing studies which help reveal fundamental properties of the plasma chemistry and its effects.

Author Contributions: Conceptualisation, Y.G.; Writing—original draft preparation, Y.G., V.S.-v.d.G.; Writing—review and editing, Y.G., J.G., V.S.-v.d.G., A.B.; Supervision, V.S.-v.d.G., A.B.

Funding: J.G. and V.S.-v.d.G. gratefully acknowledge funding by the DFG within PAK 816 "Plasma-cell interactions in Dermatology" and CRC 1316 "Transient Atmospheric Plasmas: from plasmas to liquids to solids".

Acknowledgments: We would like to thank Deborah O'Connell (York Plasma Institute, Department of Physics, University of York, United Kingdom) and Angela Privat-Maldonado (PLASMANT, University of Antwerp) for useful discussions.

Conflicts of Interest: The authors declare no conflict of interest.

References

1. Adamovich, I.; Baalrud, S.D.; Bogaerts, A.; Bruggeman, P.J.; Cappelli, M.; Colombo, V.; Czarnetzki, U.; Ebert, U.; Eden, J.G.; Favia, P.; et al. The 2017 Plasma Roadmap: Low temperature plasma science and technology. *J. Phys. D Appl. Phys.* **2017**, *50*, 323001. [CrossRef]
2. Nikiforov, A.; Chen, Z. (Eds.) *Atmospheric Pressure Plasma—From Diagnostics to Applications*; InTechOpen: London, UK, 2019.
3. Bruggeman, P.J.; Kushner, M.J.; Locke, B.R.; Gardeniers, J.G.E.; Graham, W.G.; Graves, D.B.; Hofman-Caris, R.C.H.M.; Maric, D.; Reid, J.P.; Ceriani, E.; et al. Plasma-liquid interactions: A review and roadmap. *Plasma Sources Sci. Technol.* **2016**, *25*, 053002. [CrossRef]
4. Wardenier, N.; Gorbanev, Y.; Van Moer, I.; Nikiforov, A.; Van Hulle, S.; Surmont, P.; Lynen, F.; Leys, C.; Bogaerts, A.; Vanraes, P. Removal of alachlor in water by non-thermal plasma: Reactive species and pathways in batch and continuous process. *Water Res.* **2019**. [CrossRef] [PubMed]
5. Bogaerts, A.; Neyts, E.C. Plasma Technology: An Emerging Technology for Energy Storage. *ACS Energy Lett.* **2018**, *3*, 1013–1027. [CrossRef]

6. Laroussi, M.; Akan, T. Arc-Free Atmospheric Pressure Cold Plasma Jets: A Review. *Plasma Process. Polym.* **2007**, *4*, 777–788. [CrossRef]
7. Winter, J.; Brandenburg, R.; Weltmann, K.-D. Atmospheric pressure plasma jets: An overview of devices and new directions. *Plasma Sources Sci. Technol.* **2015**, *24*, 064001. [CrossRef]
8. Lu, X.; Naidis, G.V.; Laroussi, M.; Reuter, S.; Graves, D.B.; Ostrikov, K. Reactive species in non-equilibrium atmospheric-pressure plasmas: Generation, transport, and biological effects. *Phys. Rep.* **2016**, *630*, 1–84. [CrossRef]
9. Laroussi, M. Plasma Medicine: A Brief Introduction. *Plasma* **2018**, *1*, 47–60. [CrossRef]
10. Gorbanev, Y.; O'Connell, D.; Chechik, V. Non-Thermal Plasma in Contact with Water: The Origin of Species. *Chem. Eur. J.* **2016**, *22*, 3496–3505. [CrossRef]
11. Reuter, S.; von Woedtke, T.; Weltmann, K.-D. The kINPen—A review on physics and chemistry of the atmospheric pressure plasma jet and its applications. *J. Phys. D Appl. Phys.* **2018**, *51*, 233001. [CrossRef]
12. Attri, P.; Kumar, N.; Park, J.H.; Yadav, D.K.; Choi, S.; Uhm, H.S.; Kim, I.T.; Choi, F.H.; Lee, W. Influence of reactive species on the modification of biomolecules generated from the soft plasma. *Sci. Rep.* **2015**, *5*, 8221. [CrossRef] [PubMed]
13. Gorbanev, Y.; Privat-Maldonado, A.; Bogaerts, A. Analysis of short-lived reactive species in plasma-air-water systems: The dos and the do nots. *Anal. Chem.* **2018**, *90*, 13151–13158. [CrossRef] [PubMed]
14. Walsh, J.L.; Kong, M.G. Contrasting characteristics of linear-field and cross-field atmospheric plasma jets. *Appl. Phys. Lett.* **2008**, *93*, 111501. [CrossRef]
15. MP1101—Biomedical Applications of Atmospheric Pressure Plasma Technology, European Cooperation in Science and Technology (COST). Available online: https://www.cost.eu/actions/MP1101 (accessed on 30 June 2019).
16. Knake, N.; Schulz-von der Gathen, V. Investigations of the spatio-temporal build-up of atomic oxygen inside the micro-scaled atmospheric pressure plasma jet. *Eur. Phys. J. D* **2010**, *60*, 645–652. [CrossRef]
17. Hemke, T.; Wollny, A.; Gebhardt, M.; Brinkmann, R.P.; Mussenbrock, T. Spatially resolved simulation of a radio-frequency driven micro-atmospheric pressure plasma jet and its effluent. *J. Phys. D Appl. Phys.* **2010**, *44*, 285206. [CrossRef]
18. Golda, J.; Held, J.; Redeker, B.; Konkowski, M.; Beijer, P.; Sobota, A.; Kroesen, G.; Braithwaite, N.S.J.; Reuter, S.; Turner, M.M.; et al. Concepts and characteristics of the 'COST Reference Microplasma Jet'. *J. Phys. D Appl. Phys.* **2016**, *49*, 08400. [CrossRef]
19. Schneider, S.; Jarzina, F.; Lackmann, J.-W.; Golda, J.; Layes, V.; Schulz-von der Gathen, V.; Bandow, J.E.; Benedikt, J. Summarizing results on the performance of a selective set of atmospheric plasma jets for separation of photons and reactive particles. *J. Phys. D Appl. Phys.* **2015**, *48*, 444001. [CrossRef]
20. Murakami, T.; Niemi, K.; Gans, T.; O'Connell, D.; Graham, W.G. Interacting kinetics of neutral and ionic species in an atmospheric-pressure helium–oxygen plasma with humid air impurities. *Plasma Sources Sci. Technol.* **2013**, *22*, 045010. [CrossRef]
21. Murakami, T.; Niemi, K.; Gans, T.; O'Connell, D.; Graham, W.G. Afterglow chemistry of atmospheric-pressure helium–oxygen plasmas with humid air impurity. *Plasma Sources Sci. Technol.* **2014**, *23*, 025005. [CrossRef]
22. Schröter, S.; Wijaikhum, A.; Gibson, A.R.; West, A.; Davies, H.L.; Minesi, N.; Dedrick, J.; Wagenaars, E.; de Oliveira, N.; Nahon, L.; et al. Chemical kinetics in an atmospheric pressure helium plasma containing humidity. *Phys. Chem. Chem. Phys.* **2018**, *20*, 24263–24286. [CrossRef]
23. Dedrick, J.; Schröter, S.; Niemi, K.; Wijaikhum, A.; Wagenaars, E.; de Oliveira, N.; Nahon, L.; Booth, J.-P.; O'Connell, D.; Gans, T. Controlled production of atomic oxygen and nitrogen in a pulsed radio-frequency atmospheric-pressure plasma. *J. Phys. D Appl. Phys.* **2017**, *50*, 455204. [CrossRef]
24. Ellerweg, D.; von Keudell, A.; Benedikt, J. Unexpected O and O_3 production in the effluent of He/O_2 microplasma jets emanating into ambient air. *Plasma Sources Sci. Technol.* **2012**, *21*, 034019. [CrossRef]
25. Gorbanev, Y.; Van der Paal, J.; Van Boxem, W.; Dewilde, S.; Bogaerts, A. Reaction of chloride anion with atomic oxygen in aqueous solutions: Can cold plasma help in chemistry research? *Phys. Chem. Chem. Phys.* **2019**, *21*, 4117–4121. [CrossRef] [PubMed]
26. Oehrlein, G.S.; Phaneuf, R.J.; Graves, D.B. Plasma-polymer interactions: A review of progress in understanding polymer resist mask durability during plasma etching for nanoscale fabrication. *J. Vac. Sci. Technol. B* **2011**, *29*, 010801. [CrossRef]

27. West, A.; van der Schans, M.; Xu, C.; Cooke, M.; Wagenaars, E. Fast, downstream removal of photoresist using reactive oxygen species from the effluent of an atmospheric pressure plasma jet. *Plasma Sources Sci. Technol.* **2016**, *25*, 02LT01. [CrossRef]

28. Hefny, M.M.; Nečas, D.; Zajíčková, L.; Benedikt, J. The transport and surface reactivity of O atoms during the atmospheric plasma etching of hydrogenated amorphous carbon films. *Plasma Sources Sci. Technol.* **2019**, *28*, 035010. [CrossRef]

29. Rezaei, F.; Gorbanev, Y.; Chys, M.; Nikiforov, A.; Van Hulle, S.W.H.; Cos, P.; Bogaerts, A.; De Geyter, N. Investigation of plasma-induced chemistry in organic solutions for enhanced electrospun PLA nanofibers. *Plasma Process. Polym.* **2018**, *15*, 1700226. [CrossRef]

30. Iqbal, M.; Dinh, D.K.; Abbas, Q.; Imran, M.; Sattar, H.; Ahmad, A.U. Controlled Surface Wettability by Plasma Polymer Surface Modification. *Surfaces* **2019**, *2*, 349–371. [CrossRef]

31. Shaw, D.; West, A.; Bredin, J.; Wagenaars, E. Mechanisms behind surface modification of polypropylene film using an atmospheric-pressure plasma jet. *Plasma Sources Sci. Technol.* **2016**, *25*, 065018. [CrossRef]

32. Reuter, R.; Ellerweg, D.; von Keudell, A.; Benedikt, J. Surface reactions as carbon removal mechanism in deposition of silicon dioxide films at atmospheric pressure. *Appl. Phys. Lett.* **2011**, *98*, 111502. [CrossRef]

33. Kasuya, M.; Yasui, S.; Noda, M. Deposition of SiO_2 Thin Films on Polycarbonate by Atmospheric-Pressure Plasma. *Jpn. J. Appl. Phys.* **2012**, *51*, 01AC01. [CrossRef]

34. Reuter, R.; Rügner, K.; Ellerweg, D.; de los Arcos, T.; von Keudell, A.; Benedikt, J. The Role of Oxygen and Surface Reactions in the Deposition of Silicon Oxide Like Films from HMDSO at Atmospheric Pressure. *Plasma Process. Polym.* **2012**, *9*, 1116–1124. [CrossRef]

35. Rügner, K.; Reuter, R.; Ellerweg, D.; de los Arcos, T.; von Keudell, A.; Benedikt, J. Insight into the Reaction Scheme of SiO_2 Film Deposition at Atmospheric Pressure. *Plasma Process. Polym.* **2013**, *10*, 1061–1073. [CrossRef]

36. Hefny, M.M.; Pattyn, C.; Lukes, P.; Benedikt, J. Atmospheric plasma generates oxygen atoms as oxidizing species in aqueous solutions. *J. Phys. D Appl. Phys.* **2016**, *49*, 404002. [CrossRef]

37. Benedikt, J.; Hefny, M.M.; Shaw, A.; Buckley, B.R.; Iza, F.; Schäkermann, S.; Bandow, J.E. The fate of plasma-generated oxygen atoms in aqueous solutions: Non-equilibrium atmospheric pressure plasmas as an efficient source of atomic $O_{(aq)}$. *Phys. Chem. Chem. Phys.* **2018**, *20*, 12037–12042. [CrossRef] [PubMed]

38. Lin, A.; Gorbanev, Y.; De Backer, J.; Van Loenhout, J.; Van Boxem, W.; Lemière, F.; Cos, P.; Dewilde, S.; Smits, E.; Bogaerts, A. Non-Thermal Plasma as a Unique Delivery System of Short-Lived Reactive Oxygen and Nitrogen Species for Immunogenic Cell Death in Melanoma Cells. *Adv. Sci.* **2019**, *6*, 1802062. [CrossRef] [PubMed]

39. Kondeti, V.S.S.K.; Phan, C.Q.; Wende, K.; Jablonowski, H.; Gangal, U.; Granick, J.L.; Hunter, R.C.; Bruggeman, P.J. Long-lived and short-lived reactive species produced by a cold atmospheric pressure plasma jet for the inactivation of Pseudomonas aeruginosa and Staphylococcus aureus. *Free Radic. Biol. Med.* **2018**, *124*, 275–287. [CrossRef] [PubMed]

40. Jirásek, V.; Lukeš, P. Formation of reactive chlorine species in saline solution treated by non-equilibrium atmospheric pressure He/O_2 plasma jet. *Plasma Sources Sci. Technol.* **2019**, *28*, 035015. [CrossRef]

41. Guo, Z.; Zhou, C.; Hu, S.; Chen, Y.; Jia, X.; Lau, R.; Yang, Y. Epoxidation of *trans*-stilbene and *cis*-cyclooctene over mesoporous vanadium catalysts: Support composition and pore structure effect. *Appl. Catal. A Gen.* **2012**, *419*, 194–202. [CrossRef]

42. Gorbanev, Y.; Leifert, D.; Studer, A.; O'Connell, D.; Chechik, V. Initiating radical reactions with non-thermal plasmas. *Chem. Commun.* **2017**, *53*, 3685–3688. [CrossRef]

43. Iza, F. Plasma-Driven Organic Synthesis: Waste-Free Epoxidation. In Proceedings of the 24th International Symposium on Plasma Chemistry, Naples, Italy, 9–14 July 2019.

44. Privat-Maldonado, A.; Gorbanev, Y.; O'Connell, D.; Vann, R.; Chechik, V.; van der Woude, M.W. Non-target biomolecules alter macromolecular changes induced by bactericidal low-temperature plasma. *IEEE Trans. Radiat. Plasma Med. Sci.* **2018**, *2*, 121–128. [CrossRef] [PubMed]

45. Yan, D.; Sherman, J.H.; Keidar, M. Cold atmospheric plasma, a novel promising anti-cancer treatment modality. *Oncotarget* **2017**, *8*, 15977–15995. [CrossRef] [PubMed]

46. Graves, D.B. Mechanisms of Plasma Medicine: Coupling Plasma Physics, Biochemistry, and Biology. *IEEE Trans. Radiat. Plasma Med. Sci.* **2017**, *1*, 281–292. [CrossRef]

47. Gorbanev, Y.; Soriano, R.; O'Connell, D.; Chechik, V. An atmospheric pressure plasma setup to investigate the reactive species formation. *J. Vis. Exp.* **2016**, *117*, e54765. [CrossRef] [PubMed]

48. Gorbanev, Y.; Verlackt, C.C.W.; Tinck, S.; Tuenter, E.; Foubert, K.; Cos, P.; Bogaerts, A. Combining experimental and modelling approaches to study the sources of reactive species induced in water by the COST RF plasma jet. *Phys. Chem. Chem. Phys.* **2018**, *20*, 2797–2808. [CrossRef]

49. Lackmann, J.-W.; Schneider, S.; Edengeiser, E.; Jarzina, F.; Brinckmann, S.; Steinborn, E.; Havenith, M.; Benedikt, J.; Bandow, J.E. Photons and particles emitted from cold atmospheric-pressure plasma inactivate bacteria and biomolecules independently and synergistically. *J. R. Soc. Interface* **2013**, *10*, 20130591. [CrossRef]

50. Lackmann, J.-W.; Wende, K.; Verlackt, C.; Golda, J.; Volzke, J.; Kogelheide, F.; Held, J.; Bekeschus, S.; Bogaerts, A.; Schulz-von der Gathen, V.; et al. Chemical fingerprints of cold physical plasmas—An experimental and computational study using cysteine as tracer compound. *Sci. Rep.* **2018**, *8*, 7736. [CrossRef]

51. Vermeylen, S.; De Waele, J.; Vanuytsel, S.; De Backer, J.; Van der Paal, J.; Ramakers, M.; Leyssens, K.; Marcq, E.; Van Audenaerde, J.; Smits, E.L.J.; et al. Cold atmospheric plasma treatment of melanoma and glioblastoma cancer cells. *Plasma Process. Polym.* **2016**, *13*, 1195–1205. [CrossRef]

52. Van der Paal, J.; Neyts, E.C.; Verlackt, C.C.W.; Bogaerts, A. Effect of lipid peroxidation on membrane permeability of cancer and normal cells subjected to oxidative stress. *Chem. Sci.* **2016**, *7*, 489–498. [CrossRef]

53. Bekeschus, S.; Wende, K.; Hefny, M.M.; Rödder, K.; Jablonowski, H.; Schmidt, A.; von Woedtke, T.; Weltmann, K.-D.; Benedikt, J. Oxygen atoms are critical in rendering THP-1 leukaemia cells susceptible to cold physical plasma-induced apoptosis. *Sci. Rep.* **2016**, *7*, 2791. [CrossRef]

54. Keidar, M. A prospectus on innovations in the plasma treatment of cancer. *Phys. Plasmas* **2018**, *25*, 083504. [CrossRef]

55. Heirman, P.; Van Boxem, W.; Bogaerts, A. Reactivity and stability of plasma-generated oxygen and nitrogen species in buffered water solution: A computational study. *Phys. Chem. Chem. Phys.* **2019**, *21*, 12881–12894. [CrossRef] [PubMed]

56. Privat-Maldonado, A.; Gorbanev, Y.; Dewilde, S.; Smits, E.; Bogaerts, A. Reduction of human glioblastoma spheroids using cold atmospheric plasma: The combined effect of short- and long-lived reactive species. *Cancers* **2018**, *10*, 394. [CrossRef] [PubMed]

 *plasma*

Article

Investigation of Power Transmission of a Helium Plasma Jet to Different Dielectric Targets Considering Operating Modes

Tilman Teschner [1,2,3], Robert Bansemer [1], Klaus-Dieter Weltmann [1] and Torsten Gerling [1,2,3,*]

1   Leibniz Institute for Plasma Science and Technology (INP Greifswald), Felix-Hausdorff-Str. 2, 17489 Greifswald, Germany
2   Centre for Innovation Competence (ZIK plasmatis), Felix-Hausdorff-Str. 2, 17489 Greifswald, Germany
3   Competency Center Diabetes Karlsburg (KDK), Greifswalder Str. 11, 17495 Karlsburg, Germany
*   Correspondence: gerling@inp-greifswald.de

Received: 12 April 2019; Accepted: 12 August 2019; Published: 22 August 2019

Abstract: The interaction of an atmospheric pressure plasma jet with different dielectric surfaces is investigated using a setup of two ring electrodes around a ceramic capillary. In this study, in addition to electrical measurement methods such as the determination of voltage and current, special emphasis was placed on the power measurements at the electrodes and the effluent. The power dissipation is correlated with Fourier-transform infrared (FTIR) absorption spectroscopy measurements of O_3 and NO_2 densities. The results show the correlation between the dielectric constant and the dissipated power at the target. The ratio between power dissipation at the grounded ring electrode and the grounded surface shows an increase with increasing dielectric constant of the target. A correlation of the results with bacteria, tissue and water as envisaged dielectric targets shows four times the power dissipation at the treatment spot between bacteria and tissue.

Keywords: power dissipation; plasma diagnostics; atmospheric pressure plasma jet; plasma medicine; dielectric surface; dielectric properties; Lissajous figure; operation mode

---

1. Introduction

Plasma devices operating at atmospheric pressure are a useful tool for many applications, from exhaust treatment to medical use [1–3]. The unique properties of the devices arise from the local generation of multiple reactive species on the spot [4]. A broad range of investigations consider the application of plasma treatment of surfaces aiming to coat, decontaminate or heal specific surfaces. However, the diagnostics of the devices were performed while operating without a target in front [5–7].

One major application for plasma being researched is the field of plasma medicine. In plasma medicine, cold atmospheric pressure plasma is found to inactivate a broad spectrum of microorganisms in wounds and to stimulate cell proliferation and tissue regeneration mediated by direct treatment via ultraviolet radiation and creation of reactive species or by indirect effects through excitation of the liquid phase of the cell or wound [8].

For plasma setups, the characteristics like emission, species generation, electrical field strength amplitude and even stability are impacted when a surface is put in close vicinity [9–14]. Thus, an increased interest has arisen in recent years in the investigation of interaction with a target made of liquid, dielectric or metal resembling specific application conditions [15–19]. The range of dielectric constants varies considerably ranging from low values in alumina to virtually infinity for metals.

The present contribution provides a systematic investigation on the effect of dielectric properties of different surfaces in front of a plasma jet on the discharge stability and power distribution. A plasma jet with a dielectric capillary and two ring electrodes (high voltage and grounded) as well as a grounded electrode behind the dielectric surface is investigated with power measurements at both electrodes for a variety of six different dielectric surfaces ($\epsilon_r = 2.25\ldots160$). The power consumption at the ring electrode and the surface electrode is measured for all six dielectrics as well as for different applied voltages. The distribution of power is derived from these measurements. Density measurements of O_3 and NO_2 by Fourier-transform infrared absorption spectroscopy allow an insight of the target permittivity on species production efficiency. Finally, the influence of permittivity on power and species development in plasma is used to get qualitative predictions on medically relevant targets such as *E. coli*, tissue or water.

2. Materials and Methods

2.1. Experimental Setup

Figure 1 shows a schematic of the experimental setup. The plasma source consisted of an alumina capillary with a 3 mm outer and a 1.1 mm inner diameter. A gas flow of 2 slm helium was introduced through the capillary. A sinusoidal voltage U_{app} was applied on the first copper electrode (high voltage (HV) electrode in Figure 1). To create a high voltage, a sinusoidal signal generated by a waveform generator (f = 17.9 kHz, PicoScope 3460B, Pico Technology, St Neots, UK) was amplified by a power amplifier (AG 1021, T&C Power, Rochester, NY, USA) in series with a high-voltage transformer. The second copper electrode was grounded. The dielectric surface had a distance d_T to the opposing capillary orifice and was grounded via an attached copper plate. A high voltage probe V_0 (P6015A, Tektronix, Inc. Beaverton, OR, USA) was connected to the first ring electrode for high voltage measurements. To measure the input power, a current monitor (6585 Pearson Electronics, Palo Alto, CA, USA) was placed around the input HV line. Optionally, the current over $R_i = 100\ \Omega$ resistors or the voltage drop over Styroflex capacitors $C_i = 220$ pF was measured with voltage probes V_i (TA131, Pico Technology, St Neots, UK), where i represents the grounded ring electrode (1) and the grounded counter electrode (2).

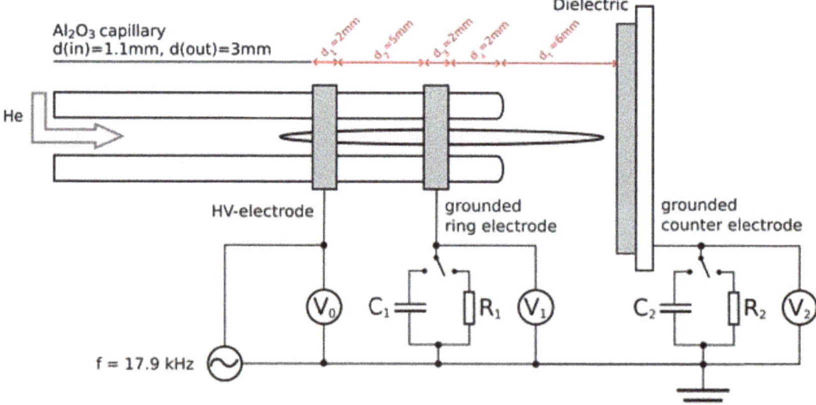

Figure 1. Experimental setup of the discharge device, including the electrical circuit and the dimensions of the electrode placement. The variables in this study are the applied voltage U_{app} and the dielectric constant ϵ_r of the external dielectric surface.

The dielectrics used in this experiment with their relative permittivity and their major associated polarization mechanisms are shown in Table 1. Values for water, tissue and *E. coli* are added as reference [20–22] due to their relevance in plasma medicine.

Table 1. Dielectric materials used in this study with dielectric constant, polarization effects and sample values for water, tissue and *E. coli*. EP—electronic polarization, IP—ionic polarization, DP—dipolar polarization [20]. *** indicates no available information on polarization effects.

Dielectric	PE	SiO$_2$	B270	Al$_2$O$_3$	ZrO$_2$	TiO$_2$	H$_2$O	Tissue	*E. coli*
Permittivity	2.25	4.3	7,0	9,0	22	160	80	60	6.5
Polarization	EP, IP	EP, IP	EP, IP	EP, IP, DP	EP, IP, DP	EP, IP, DP	EP, IP, DP	***	***

2.2. Power Determination

The transferred charges could be determined from the measured charging curves of the capacitors. Relating these charges with the input voltages, one received Lissajous figures. The power dissipation could be determined from the area of these figures. Due to the setup with three electrodes, nonlinear plasma dynamics arose, so that the classical power determination via the Manley equation [23] was not possible. Charge transfer and discharges between the electrodes caused the Lissajous figure to fold, which could result in intersections of the curve. Using the Gauss's area formula, the enclosed area A_P of the resulting figure can be calculated with the vertices (x_j, y_j) [24]:

$$A_P = \frac{1}{2} \left| \sum_{j=0}^{n} -1(y_j + y_{j+1 \bmod n})(x_j - x_{j+1 \bmod n}) \right|. \tag{1}$$

This area corresponded to the dissipated energy E_{dis} of the discharges. By including the frequency, the mean dissipated power P_i could be determined with

$$P_i = \frac{1}{T} \int_0^T U_{\mathrm{app}}(t) \cdot C_i \frac{dU_i(t)}{dt} = f \underbrace{\oint_{\substack{\mathrm{one} \\ \mathrm{cycle}}} U_{\mathrm{app}} \, dQ_i}_{=E_{\mathrm{dis}}=A_P} \tag{2}$$

for one of the electrodes $i = 1, 2$ considering the applied voltage U_{app} [25]. In this geometry, P_i describes the power dissipated at the respective electrode.

The input power P_{in} was acquired by measuring U_{app} and the input current I_0. To consider power losses, the power measurement was performed also without plasma ignition P_0 by switching off the helium flow. The calculations were comparable with [6]:

$$P_{in/0} = \frac{1}{T} \int_0^T U_{\mathrm{app}}(t) \cdot I_0(t) dt \ (\text{plasma on}/\text{off}), \tag{3}$$

$$P_{\mathrm{plasma}} = P_{\mathrm{in}} - P_0. \tag{4}$$

Finally, the power dissipated into the effluent was described as the loss into emission and species production and was determined by

$$P_{\mathrm{effluent}} = P_{\mathrm{plasma}} - P_1 - P_2. \tag{5}$$

2.3. Determination of Produced Species Densities

In order to obtain an insight into the plasma chemical effects triggered by exchanging the treated dielectric and by the concomitant change in electrical power dissipation, absolute densities of ozone (O_3) and nitrogen dioxide (NO_2) were measured using Fourier-transform infrared (FTIR) absorption spectroscopy. A sufficient sensitivity for far-field measurements was provided by attaching a 32 m multi-pass cell to the spectrometer in use (Vertex 80v, Bruker Corporation, Billerica, MA, USA).

The desired species were expected to be produced in reactions with the surrounding atmosphere. Therefore, a reproducible environment needed to be created. This is realized by operating the plasma source in a glass cell that was flushed with 5 slm of dry, artificial air (80% N_2, 20% O_2). A vacuum pump at the multi-pass cell outlet and a throttle valve at its inlet led to a pressure of 280 mbar in the cell. Over the throttle valve, the gas produced by the plasma source in its controlled atmosphere was aspirated into the multi-pass cell. Figure 2 depicts the overall setup that has previously been used e.g., in [26].

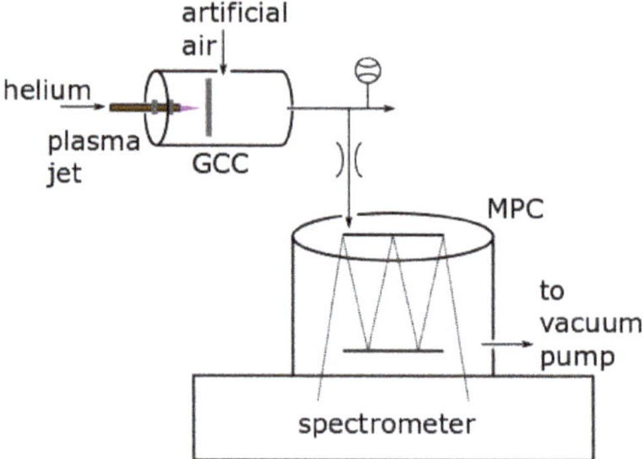

Figure 2. Fourier-transform infrared spectroscopy setup used to measure absolute densities of O_3 and NO_2. *GCC* is the gas collector cell and *MPC* is the multi-pass cell.

The measurement captured a wavenumber range of 700 to 4000 cm^{-1}, which allowed the detection of several nitrogen compounds such as nitric oxide (NO) and nitrous oxide (N_2O) in addition to the expected most abundant species O_3 and NO_2 in the far-field. Consequently, an unexpected generation of byproducts in concentrations above the qualitative detection limit around 5×10^{12} cm^{-3} can be detected as well. The limit for a sound quantitative analysis in the measurement system is in the range of 1×10^{13} cm^{-3}. Density values were calculated by fitting reference spectra from the HITRAN database [27] using the absorbance function

$$A_{FTIR} = -\ln \frac{I(\nu)}{I_0(\nu)} = \sum_i n_i \sigma_i(\nu) L. \tag{6}$$

3. Results

The discharge dynamics are examined by electrical characterization in the following section. Basic electrical characteristics of the setup shown in Figure 1 are presented in Figure 3. The observed current pulses show an asymmetric current signal that is typical for this measurement setup consisting of three electrodes [11]. Due to the applied sinusoidal voltage ($U_{app} = 9.0$ kV$_{pp}$), the generated charge carriers are separated. The ions accumulate on the surface of the cathode, the electrons on surface of the anode. As a result, an electric field is generated that counteracts the external field generated by the voltage (up to #1 in Figure 3). As the external applied voltage decreases, the internal electric field causes the discharge in the capillary. A fast current peak of $I_1 = 2.1$ mA and a duration of 1.5 µs is observed at the grounded ring electrode (#1 in Figure 3). This leads to further ionization of the helium gas, which flows through the capillary. The resulting ions partially recombine with surface charge carriers on the inner surface of the capillary. Another part of the ions is accelerated by the internal electric field towards the grounded ring electrode. These ions accumulate on the capillary inside of the grounded ring electrode and partly diffuse to the capillary edge, which results in the formation of a plasma bullet. The bullet is accelerated by the electric field of the surface charge carriers towards the grounded counter electrode and impinge on the surface of the dielectric target. A current pulse of inverse polarity with a peak value of $I_1 = 0.25$ mA and a duration of 1 µs is measured (#2 in Figure 3).

A fast 'return stroke', consisting of electrons, leads from the surface of the counter electrode to the capillary edge, leaving behind a charge channel for further ions [28,29].

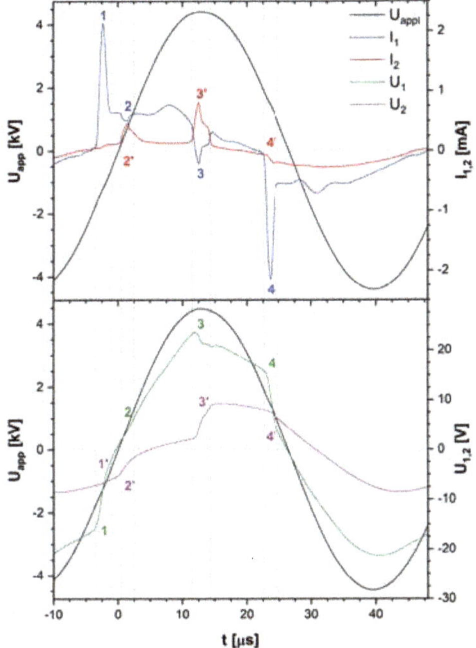

Figure 3. Current and voltage characteristics of the discharge in helium with SiO_2 as dielectric target ($U_{appl} = 9.0\,kV_{pp}$, $d_T = 6$ mm). 1–4 and 1'–4' mark characteristic events.

Due to the steadily increasing input voltage and the resulting electric field, additional ions are deposited at the capillary edge, which initiates a further discharge to the dielectric. The discharge leads to a current drop of 0.7 mA with a duration of 3 μs at the ring electrode (#3 in Figure 3). Similar to the previous discharge, a 'return stroke' leads to compensation of ions and electrons between the capillary and the dielectric. As soon as the input voltage drops, the external electric field decreases. The electric field induced by the surface charge carriers leads to the acceleration of the electrons from the HV-electrode towards the ring electrode. A negative current peak of $I_1 = -2.25$ mA and a duration of 1.5 μs is observed at the grounded ring electrode (#4 in Figure 3). The ions accelerate in the opposite direction. The electrons are further accelerated to the counter electrode leaving a charge channel, so that a backward bullet can be formed. Due to the negative current signal between 25 μs and 45 μs electrons diffuse to the counter electrode and restore the starting conditions [30].

The comparison of the discharge curves of the capacitors shows the charge transport across the plasma column (see Figure 3). The discharges within the capillary cause a change in the slope of the measured voltage U_1 (#1, #4 in Figure 3). A major change in voltage U_2 is observed when the bullet strikes the dielectric (#2', #3' in Figure 3). The charge transport is the key parameter for the power dissipated in the plasma P_1 and the power transmission to the counter electrode P_2.

The power is determined by the enclosed area of the Lissajous figures calculated with the Gauss's area formula (see Equation (1)). Examples of Lissajous figures for the grounded ring electrode using different dielectrics are shown in Figure 4. Lissajous figures on the counter electrode have a rounded rectangular shape (not shown here). Similar to the current signal, the voltage value is asymmetric due to the setup with three electrodes, which results in charge interaction between the dielectric and the capillary of the plasma source. As a result, nodal points can occur in the Lissajous figures for our setup.

Therefore, the relative permittivity is a decisive influencing factor. While one obtains a continuous area for the Lissajous figure when applying the plasma to PE ($\epsilon_{r,PE} = 2.25$), the number of nodes increases with increasing dielectric constant ($\epsilon_{r,ZrO_2} = 22$, $\epsilon_{r,TiO_2} = 160$). Each node correlates with a charge exchange process taking place and charges that are stored on the capillary being delivered to the counter electrode. With the increased number of nodes for increasing ϵ_r, a decrease of the encircled area is observed, indicating a reduction in power at the ring electrode. When evaluating the curves in Figure 4 that contain nodes, one has to consider that automated routines as in Origin (Origin2018b, OriginLab Corporation, Northampton, MA, USA) generate negative area values that would stand for negative powers. To avoid these errors, the Gauss's area formula was implemented to calculate the area under the curves.

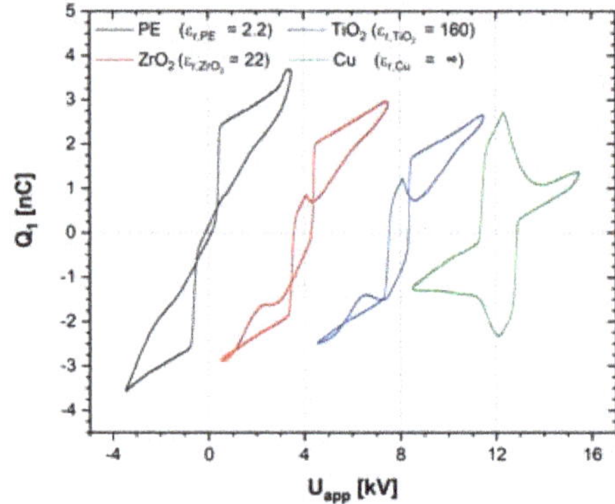

Figure 4. Q–V-plot for representative dielectric targets and copper target with an applied voltage offset $U_{app,off} = 4.0$ kV between each dielectric ($d_T = 6$ mm).

In Figure 5, the measured dissipated power at the ring electrode and the counter electrode is shown for different applied voltages. Each curve represents one dielectric target at the counter electrode. With increasing voltage, the dissipated power at the ring electrode starts to rise until it reaches 2 kV (first grey bar). Above 2 kV, a drop in dissipated power is observed for each dielectric material followed by a rise up to 4 kV (second grey bar). Around a voltage of 4 kV, most dielectrics show a second drop in dissipated power at the ring electrode that correlates with a second inverse current pulse. Only for TiO$_2$, which has the the highest ϵ_r, this second drop was not observed within the operated voltage range.

Considering the dissipated power at the counter electrode shown in Figure 5, a weak increase is observed until 2 kV. Above 2 kV, correlating with the drop of the power at the ring electrode P_1, a first strong increase in dissipated power at the counter electrode P_2 is observed. While the power at the ring electrode drops about 80 mW, the increase at the counter electrodes rises above 200 mW. With a further rise in voltage, the dissipated power increases nearly linearly again until around 4 kV; a second strong increase is observed for most dielectrics except for TiO$_2$. One further observation is the increase of U_{app} required for the second strong increase when increasing the dielectric permittivity.

A surface with a lower permittivity resembles a lower capacity and is charged within a shorter time frame compared to higher ϵ_r. Once charged until a certain point, the local electrical field of the surface charges generate a counteracting electrical field negating the external electrical field from the setup. By further increasing the applied voltage U_{app}, the external field is increased again until a further ignition is enabled considering memory effects and hence reduced ignition requirements (see #3 and #3′ in Figure 3). For higher ϵ_r, a higher capacity has to be charged and the additional charges are distributed through surface ionization waves (SIWs) [10]. As a result of the longer charging time and increased spatial distribution through SIWs, the countering of the external electrical field requires a longer time frame and ultimately a higher U_{app} for the second discharge signal.

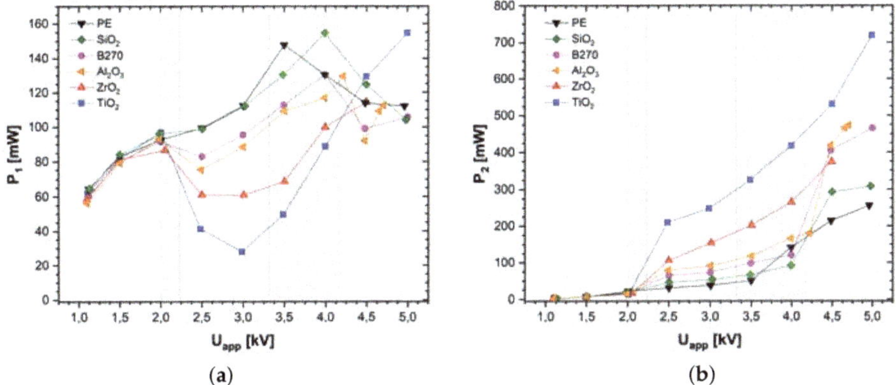

Figure 5. Dissipated power $P_i(\epsilon_r)$ for (**a**) grounded ring electrode (i = 1) and (**b**) grounded surface electrode (i = 2) at different applied voltages (d_T = 6 mm).

While the dissipated power at the counter electrode seems to rise constantly, a saturation or limitation seems to occur at the discharge inside the capillary. Our previous investigation showed a similar saturation in the 'core' discharge in a different geometry at about 1 W [6]. In the present setup, the limit is between 130 mW and 160 mW depending on the dielectric target. One interpretation is a saturation of the charge accumulation on the dielectric capillary resembling a geometrical limitation [6]. Another reason for the limitation could be the additional induced electric field by the deposited charged that generates a second discharge channel (#3′ in Figure 3). As a result, a second conductive channel for charge exchange is formed, which has been previously measured [11,30].

Now, correlating the dissipated power with the species production by measuring the far-field densities shows a clear tendency towards an increased density by up to one order of magnitude for an applied voltage of 3.5 kV compared to 2 kV (Figure 6). For $U_{app,off}$ = 2 kV, species densities of O_3 and NO_2 do not exceed the quantitative detection limit of about 1×10^{13} cm^{-3}, yet qualitative detection was possible in all cases (low densities). For $U_{app,off}$ = 3.5 kV, the O_3 density increases with the increasing permittivity from 1×10^{13} cm^{-3} up to 4×10^{13} cm^{-3} while NO_2 does not reach a quantitative detection limit of 1×10^{13} cm^{-3} but was observed qualitatively again. Other species absorbing in the considered wavenumber range of 700 to 4000 cm^{-1} could not be detected.

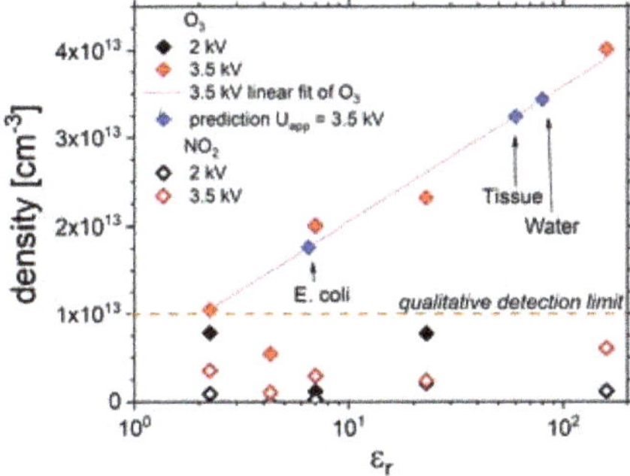

Figure 6. Measured O_3 (filled data points) and NO_2 (open data points) densities for each dielectric target and voltage amplitudes of 2 kV (black points) and 3.5 kV (red points), d_T = 6 mm. The qualitative detection limit of the species fit is inserted at 1×10^{13} cm^{-3}. The fit includes only values above the qualitative detection limit.

Since the measurement is performed in the far-field post reactions of the produced species by e.g., working gas humidity, chamber humidity or flushing gas have to be considered when discussing the total densities. This humidity is a function of time and hence not fully negligible [31]. In the present setup, it was shown previously that humidity has a negligible effect on ozone densities in the FTIR measurements while NO_2 densities might even increase with higher humidity [32].

Switching between a discharge with and without contact to the dielectric surface resembles the observation described for a different setup, where the two options are called low-power mode and high-power mode [33]. In our setup, the high-power mode seems stable and reproducible for an applied voltage of 2 kV and the low-power mode for values below 2 kV. Just above 2 kV, a phase transition occurs with a discharge towards the surface electrode in successive every fourth, third and second period before stabilizing for a discharge in every period. Another study called these modes bullet mode for in our case below 2 kV and continuous mode above 2 kV [34]. The chaotic mode was not observed in the present setup under our operating conditions despite similarities to literature [35]. However, initial indications of this mode were observed for lower helium supply quality, especially at higher voltages.

4. Discussion

In Figure 7, the power ratio of P_2 to P_1 is plotted against the dielectric permittivity of the surface for the two grey areas shown in Figure 5. The black triangles show the ratio for an applied voltage of 2 kV and the red triangles for 3.5 kV. From the power ratio with increasing permittivity, a clear tendency is observed for U_{app} = 3.5 kV, while, at U_{app} = 2 kV, the ratio stays constantly below 0.5; it increases steadily for U_{app} = 3.5 kV with increasing ϵ_r. The highest measured ratio is around 6.5 for TiO_2, hence over six times more power is dissipated outside the capillary then inside.

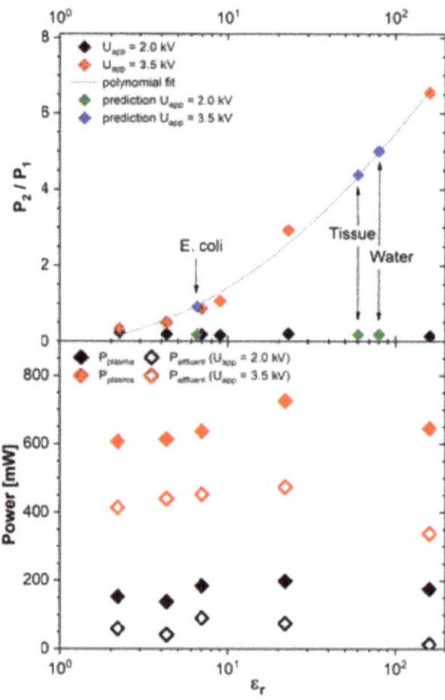

Figure 7. (**Top**) ratio of dissipated power at the target electrode (2) and the ring electrode (1) for voltage amplitudes of 2 kV and 3.5 kV. In addition, the expected values for a target electrode consisting of *E. coli*, tissue or water are extrapolated. (**Bottom**) the plasma power and the effluent power from Equations (4) and (5) are presented for voltage amplitudes of 2 kV and 3.5 kV, d_T = 6 mm.

The different ratios show the influence of the dielectric permittivity of the target on the distribution of power between the inside and outside of the capillary. For any application, this could impact the effectivity of the device, since an increase or decrease in power dissipation scales with the production of e.g., ions, emission or reactive species (see Figure 6) and thereby the impact on a surface. The effect of this feature on plasma parameters like electron temperature is an exciting topic for future investigations with this setup. First, results on fundamental plasma parameters show no significant impact in the gap but rather near and on the surface [15]. In addition, modeling results indicate an order of magnitude increase in reactive species densities in the gap [10]. In the far-field, however, we measured an increase of half an order of magnitude O_3 density.

While the overall ratio of dissipated power at the grounded counter electrode P_2 to the grounded ring electrode, P_1 grows for increasing dielectric constant of the surface (Figure 7), the input plasma power P_{plasma} remains roughly constant for the investigated range of dielectric materials. However, the effluent power $P_{effluent}$ drops visibly for TiO_2. If the identical setup is operated with a blank copper electrode as grounded counter electrode P_{plasma} is doubled, a similar P_2/P_1 ratio for TiO_2 is recorded. The power dissipation onto a copper target should be most efficient ($\epsilon_r = \infty$), while the dissipation into the effluent is still within the range of all investigated dielectric targets.

In Table 1, we noted the values for the dielectric permittivity. Considering Figure 7 and the impact of the permittivity, the value itself has to be questioned. For lower permittivities, the values differ only slightly in literature (e.g., quartz from 3 to 6), while, for ZrO_2, values from distributors already vary between 20 and 29. For TiO_2, the permittivity strongly depends on the orientation and can vary from 90 up to 180. For our investigations, the exact properties of the dielectrics could not be provided by the manufacturer. Therefore, the fit presents a first interpretation to the best of our knowledge. Nonetheless, the overall tendency of an increased ratio P_2/P_1 with increasing permittivity prevails.

In addition to the investigated dielectrics, an interpolation fit was performed to pinpoint targets of high importance for the exemplary field of plasma medicine and also for coatings of different dielectric samples. By including *E. coli*, tissue and water in the graph different power ratios should be expected from 1.0 for *E. coli* (P_2 = 75 mW) up to 4.5 for tissue (P_2 = 250 mW) and 5.0 for water (P_2 = 275 mW). In addition, the interpolation for the O_3 densities shows an increase from *E. coli* with $n_{O_3} = 1.8 \times 10^{13}$ cm^{-3} to water and tissue around $n_{O_3} = 3.5 \times 10^{13}$ cm^{-3}. Hence, the application of a plasma device on a Petri dish with *E. coli* might differ from the application on a tissue sample. How much the humid surrounding of *E. coli* within a Petri dish will impact this ratio is another topic to consider in this context. The operation of the device in the low power mode will prevent this influence while reducing the performance. Standardization and comparability are topics for advancing research and application for plasma devices of the same or even different types [36,37]. At this point, the effect of a surface on a plasma device is far from understood or even controlled.

Although an extrapolation could provide values for e.g., *E. coli*, tissue and water, it has to be taken into account that this investigation has a set of parameters such as the distance between device and surface predetermined. The distance, however, is subject to constant change under application conditions, which can lead to changes between low power mode and high power mode. An increase in distance requires an adaptation of applied voltage to lock an operation mode. We hope to raise awareness in the scientific community for the impact of the surface positioning and properties on a plasma device.

5. Conclusions

The interaction of an atmospheric pressure plasma jet with different dielectric surfaces is investigated in this study. By evaluating discharge characteristics and dissipated power inside the jet P_1 and between the jet and the surface P_2, the effect of the dielectric permittivity on the plasma performance is observed for different applied voltages at a fixed distance. The evaluated voltage-charge plots revealed several nodal points unknown in literature. Measuring the power P_1 and P_2 while increasing the voltage revealed two different operation modes described in the literature as

- low power mode—the discharge ignites mainly inside the capillary not touching the surface
- high power mode—the discharge reaches the surface, a return stroke and a secondary channel are created.

While the low power mode has no power dissipation at the surface (P_2 around zero), the high power mode dissipates power at the dielectric surface ($P_2 = 50$ mW to 300 mW). It was observed that the amplitude of P_2 rises with the applied voltage U_{app} and the permittivity ϵ_r, while P_1 increases as well but not as intensely as P_2. The increase of power dissipation indicates an increase of energy, since the time characteristics stay constant. For the dissipated power P_1, a saturation at 160 mW is indicated, which is interpreted as a geometrical dissipation restriction. In addition, O_3 and NO_2 were qualitatively observed for both modes, while only O_3 was quantitatively measured for the high power mode ($n_{O_3} = 1 \times 10^{13}$ cm^{-3} to 4×10^{13} cm^{-3}).

For each mode, the power ratio of P_2 to P_1 is evaluated for different ϵ_r, revealing a more efficient power dissipation at the target surface with increasing permittivity and a constant plasma power P_{plasma}. This correlates with an increase of O_3 densities from 1×10^{13} cm^{-3} up to 4×10^{13} cm^{-3}. In addition, the implication for medical application is stated by interpolating the expected power input onto a target surface consisting of either *E. coli*, tissue or water. In an application case, a target change from *E. coli* to tissue would result in an increase in power dissipation of four times when operating the device in the high power mode.

Author Contributions: Conceptualization, T.T. and T.G.; methodology, T.T. and R.B.; software, T.T. and R.B.; validation, T.T. and T.G.; formal analysis, T.T.; investigation, T.T. and T.G.; resources, T.G. and K.-D.W.; data curation, T.T.; writing—original draft preparation, T.T. and T.G.; writing—review and editing, T.T. and T.G.; visualization, T.G. and K.-D.W.; supervision, K.-D.W.; project administration, K.-D.W.; funding acquisition, T.G. and K.-D.W.

Funding: The authors acknowledge the funding from the Ministry of Education, Science and Culture of the State of Mecklenburg-Western Pomerania and European Union, European Social Fund, Grant Nos. AU 11 038; ESF/IV-BM-B35-0010/13; AU 15 001.

Acknowledgments: The authors acknowledge the technical assistance from Peter Holtz and Christiane Meyer as well as Ronny Brandenburg for fruitful discussions.

Conflicts of Interest: The authors declare no conflict of interest. The funders had no role in the design of the study; in the collection, analyses, or interpretation of data; in the writing of the manuscript, or in the decision to publish the results.

References

1. Bruggeman, P.; Brandenburg, R. Atmospheric pressure discharge filaments and microplasmas: Physics, chemistry and diagnostics. *J. Phys. D Appl. Phys.* **2013**, *46*, 464001. [CrossRef]
2. Winter, J.; Brandenburg, R.; Weltmann, K.D. Atmospheric pressure plasma jets: An overview of devices and new directions. *Plasma Sources Sci. Technol.* **2015**, *24*, 064001. [CrossRef]
3. Setsuhara, Y. Low-temperature atmospheric-pressure plasma sources for plasma medicine. *Arch. Biochem. Biophys.* **2016**, *605*, 3–10. [CrossRef] [PubMed]
4. Lu, X.; Naidis, G.; Laroussi, M.; Reuter, S.; Graves, D.; Ostrikov, K. Reactive species in non-equilibrium atmospheric-pressure plasmas: Generation, transport, and biological effects. *Phys. Rep.* **2016**, *630*, 1–84. [CrossRef]
5. Lu, X.; Laroussi, M.; Puech, V. On atmospheric-pressure non-equilibrium plasma jets and plasma bullets. *Plasma Sources Sci. Technol.* **2012**, *21*, 034005. [CrossRef]
6. Gerling, T.; Brandenburg, R.; Wilke, C.; Weltmann, K.D. Power measurement for an atmospheric pressure plasma jet at different frequencies: Distribution in the core plasma and the effluent. *Eur. Phys. J. Appl. Phys.* **2017**, *78*, 10801. [CrossRef]
7. Reuter, S.; von Woedtke, T.; Weltmann, K.D. The kINPen—A review on physics and chemistry of the atmospheric pressure plasma jet and its applications. *J. Phys. D Appl. Phys.* **2018**, *51*, 233001. [CrossRef]
8. Weltmann, K.D.; von Woedtke, T. Plasma medicine current state of research and medical application. *Plasma Phys. Control. Fusion* **2017**, *59*, 014031. [CrossRef]

9. Sretenović, G.B.; Krstić, I.B.; Kovačević, V.V.; Obradović, B.M.; Kuraica, M.M. Spatio-temporally resolved electric field measurements in helium plasma jet. *J. Phys. D Appl. Phys.* **2014**, *47*, 102001. [CrossRef]

10. Norberg, S.A.; Johnsen, E.; Kushner, M.J. Helium atmospheric pressure plasma jets touching dielectric and metal surfaces. *J. Appl. Phys.* **2015**, *118*, 013301. [CrossRef]

11. Gerling, T.; Wild, R.; Nastuta, A.V.; Wilke, C.; Weltmann, K.D.; Stollenwerk, L. Correlation of phase resolved current, emission and surface charge measurements in an atmospheric pressure helium jet. *Eur. Phys. J. Appl. Phys.* **2015**, *71*, 20808. [CrossRef]

12. Guaitella, O.; Sobota, A. The impingement of a kHz helium atmospheric pressure plasma jet on a dielectric surface. *J. Phys. D Appl. Phys.* **2015**, *48*, 255202. [CrossRef]

13. Kone, A.; Sainct, F.P.; Muja, C.; Caillier, B.; Guillot, P. Investigation of the Interaction between a Helium Plasma Jet and Conductive (Metal)/Non-Conductive (Dielectric) Targets. *Plasma Med.* **2017**, *7*, 333–346. [CrossRef]

14. Wang, R.; Xu, H.; Zhao, Y.; Zhu, W.; Ostrikov, K.K.; Shao, T. Effect of dielectric and conductive targets on plasma jet behaviour and thin film properties. *J. Phys. D Appl. Phys.* **2019**, *52*, 074002. [CrossRef]

15. Klarenaar, B.L.M.; Guaitella, O.; Engeln, R.; Sobota, A. How dielectric, metallic and liquid targets influence the evolution of electron properties in a pulsed He jet measured by Thomson and Raman scattering. *Plasma Sources Sci. Technol.* **2018**, *27*, 085004. [CrossRef]

16. Kovačević, V.V.; Sretenović, G.B.; Slikboer, E.; Guaitella, O.; Sobota, A.; Kuraica, M.M. The effect of liquid target on a nonthermal plasma jet—Imaging, electric fields, visualization of gas flow and optical emission spectroscopy. *J. Phys. D Appl. Phys.* **2018**, *51*, 065202. [CrossRef]

17. Slikboer, E.; Sobota, A.; Guaitella, O.; Garcia-Caurel, E. Imaging axial and radial electric field components in dielectric targets under plasma exposure. *J. Phys. D Appl. Phys.* **2018**, *51*, 115203. [CrossRef]

18. Norberg, S.A.; Parsey, G.M.; Lietz, A.M.; Johnsen, E.; Kushner, M.J. Atmospheric pressure plasma jets onto a reactive water layer over tissue: Pulse repetition rate as a control mechanism. *J. Phys. D Appl. Phys.* **2019**, *52*, 015201. [CrossRef]

19. Sobota, A.; Guaitella, O.; Sretenović, G.B.; Kovačević, V.V.; Slikboer, E.; Krstić, I.B.; Obradović, B.M.; Kuraica, M.M. Plasma-surface interaction: Dielectric and metallic targets and their influence on the electric field profile in a kHz AC-driven He plasma jet. *Plasma Sources Sci. Technol.* **2019**, *28*, 045003. [CrossRef]

20. Ivers-Tiffée, E.; von Münch, W. *Werkstoffe der Elektrotechnik*; Springer: Berlin, Germany, 2007.

21. Gabriel, S.; Lau, R.W.; Gabriel, C. The dielectric properties of biological tissues: III. Parametric models for the dielectric spectrum of tissues. *Phys. Med. Biol.* **1996**, *41*, 2271–2293. [CrossRef] [PubMed]

22. Esteban-Ferrer, D.; Edwards, M.A.; Fumagalli, L.; Juárez, A.; Gomila, G. Electric Polarization Properties of Single Bacteria Measured with Electrostatic Force Microscopy. *ACS Nano* **2014**, *8*, 9843–9849. [CrossRef] [PubMed]

23. Manley, T.C. The Electric Characteristics of the Ozonator Discharge. *Trans. Electrochem. Soc.* **1943**, *84*, 83–96. [CrossRef]

24. Braden, B. The Surveyor's Area Formula. *Coll. Math. J.* **1986**, *17*, 326–337. [CrossRef]

25. Ashpis, D.; Laun, M.; Griebeler, E. Progress toward Accurate Measurements of Power Consumption of DBD Plasma Actuators. In Proceedings of the 50th AIAA Aerospace Sciences Meeting Including the New Horizons Forum and Aerospace Exposition, Nashville, TN, USA, 9–12 January 2012.

26. Schmidt-Bleker, A.; Bansemer, R.; Reuter, S.; Weltmann, K.D. How to produce an NO_x- instead of O_x-based chemistry with a cold atmospheric plasma jet. *Plasma Process. Polym.* **2016**, *13*, 1120–1127. [CrossRef]

27. Gordon, I.E.; Rothman, L.S.; Hill, C.; Kochanov, R.V.; Tan, Y.; Bernath, P.F.; Birk, M.; Boudon, V.; Campargue, A.; Chance, K.V.; et al. The HITRAN2016 molecular spectroscopic database. *J. Quant. Spectrosc. Radiat. Transf.* **2017**, *203*, 3–69. [CrossRef]

28. Sigmond, R.S. The residual streamer channel: Return strokes and secondary streamers. *J. Appl. Phys.* **1984**, *56*, 1355–1370. [CrossRef]

29. Gerling, T.; Nastuta, A.V.; Bussiahn, R.; Kindel, E.; Weltmann, K.D. Back and forth directed plasma bullets in a helium atmospheric pressure needle-to-plane discharge with oxygen admixtures. *Plasma Sources Sci. Technol.* **2012**, *21*, 034012. [CrossRef]

30. Wild, R.; Gerling, T.; Bussiahn, R.; Weltmann, K.D.; Stollenwerk, L. Phase-resolved measurement of electric charge deposited by an atmospheric pressure plasma jet on a dielectric surface. *J. Phys. D Appl. Phys.* **2014**, *47*, 042001. [CrossRef]

31. Winter, J.; Wende, K.; Masur, K.; Iseni, S.; Dünnbier, M.; Hammer, M.U.; Tresp, H.; Weltmann, K.D.; Reuter, S. Feed gas humidity: A vital parameter affecting a cold atmospheric-pressure plasma jet and plasma-treated human skin cells. *J. Phys. D Appl. Phys.* **2013**, *46*, 295401. [CrossRef]

32. Hansen, L.; Schmidt-Bleker, A.; Bansemer, R.; Kersten, H.; Weltmann, K.D.; Reuter, S. Influence of a liquid surface on the NO_x production of a cold atmospheric pressure plasma jet. *J. Phys. D Appl. Phys.* **2018**, *51*, 474002. [CrossRef]

33. Sobota, A.; Guaitella, O.; Rousseau, A. The influence of the geometry and electrical characteristics on the formation of the atmospheric pressure plasma jet. *Plasma Sources Sci. Technol.* **2014**, *23*, 025016. [CrossRef]

34. Walsh, J.L.; Iza, F.; Janson, N.B.; Law, V.J.; Kong, M.G. Three distinct modes in a cold atmospheric pressure plasma jet. *J. Phys. D Appl. Phys.* **2010**, *43*, 075201. [CrossRef]

35. Walsh, J.L.; Iza, F.; Janson, N.B.; Kong, M.G. Chaos in atmospheric-pressure plasma jets. *Plasma Sources Sci. Technol.* **2012**, *21*, 034008. [CrossRef]

36. Golda, J.; Held, J.; Redeker, B.; Konkowski, M.; Beijer, P.; Sobota, A.; Kroesen, G.; Braithwaite, N.S.J.; Reuter, S.; Turner, M.M.; et al. Concepts and characteristics of the COST Reference Microplasma Jet. *J. Phys. D Appl. Phys.* **2016**, *49*, 084003. [CrossRef]

37. Mann, M.S.; Tiede, R.; Gavenis, K.; Daeschlein, G.; Bussiahn, R.; Weltmann, K.D.; Emmert, S.; von Woedtke, T.; Ahmed, R. Introduction to DIN-specification 91315 based on the characterization of the plasma jet kINPen® MED. *Clin. Plasma Med.* **2016**, *4*, 35–45. [CrossRef]

 *plasma*

Article

Emission Spectroscopic Characterization of a Helium Atmospheric Pressure Plasma Jet with Various Mixtures of Argon Gas in the Presence and the Absence of De-Ionized Water as a Target

Nima Bolouki [1,*] , **Jang-Hsing Hsieh [1,2]**, **Chuan Li [3] and Yi-Zheng Yang [2]**

[1]   Center for Plasma and Thin Film Technologies, Ming Chi University of Technology, New Taipei City 24301, Taiwan
[2]   Department of Materials Engineering, Ming Chi University of Technology, New Taipei City 24301, Taiwan
[3]   Department of Biomedical Engineering, National Yang Ming University, Taipei 112-21, Taiwan
*   Correspondence: bolouki@mail.mcut.edu.tw

Received: 25 April 2019; Accepted: 3 July 2019; Published: 4 July 2019

Abstract: A helium-based atmospheric pressure plasma jet (APPJ) with various flow rates of argon gas as a variable working gas was characterized by utilizing optical emission spectroscopy (OES) alongside the plasma jet. The spectroscopic characterization was performed through plasma exposure in direct and indirect interaction with and without de-ionized (DI) water. The electron density and electron temperature, which were estimated by Stark broadening of atomic hydrogen (486.1 nm) and the Boltzmann plot, were investigated as a function of the flow rate of argon gas. The spectra obtained by OES indicate that the hydroxyl concentrations reached a maximum value in the case of direct interaction with DI water as well as upstream of the plasma jet for all cases. The relative intensities of hydroxyl were optimized by changing the flow rate of argon gas.

Keywords: atmospheric pressure plasma jet; plasma characterization; optical emission spectroscopy

---

1. Introduction

Cold atmospheric plasma devices, mainly based on the atmospheric pressure plasma jet (APPJ) [1], have emerged over recent decades. Such devices offer the possibility of direct and indirect (remote) methods for bacteria inactivation [2], biofilm control [3], cancer cell treatment [4], water purification [5], plasma activated water [6], and so on. The APPJ with a tube-shaped configuration fed by helium gas was presented by Laroussi et al. [7] for the first time. Then, the APPJ was developed and extended with different electrode configurations and working gases.

Generally, an APPJ with various mixtures of feed gases provides numerous types of reactive species with different concentrations [8,9]. One of the species is a hydroxyl radical, which provides excellent benefits for biological and medical applications although it has a short lifetime [10,11]. Some articles reported the characterization of the argon-based APPJ with a mixture of water to enhance the hydroxyl species [12–14]. Argon as a working gas might be considered as an alternative candidate since scarce and costly helium is consumed in a large volume in the APPJ. The argon APPJ has better energy transfer efficiency compared to the helium jet; however, it releases considerable heat [15]. The conditions of electrical discharges in a pure argon gas might be unstable as a result of changing the parameters related to experimental conditions. For example, an additive gas such as oxygen is able to extinguish the argon discharge and shrink the range of discharge stabilities due to low electron temperature compared to the helium discharge [15,16]. Therefore, a plan of mixing argon and helium gases is proposed not only to reduce helium consumption but also to enhance the concentration of reactive species, specifically hydroxyl.

Based on the specific applications, the APPJ-based devices are designed and constructed with different power sources, geometries, and working gases, which results in changing the electrical conditions of the APPJ as well as the produced reactive species. Moreover, interactions of the plasma jet with liquid targets, which are most of the biological cases, influence the plasma conditions and the produced reactive species. Therefore, it is undoubtedly necessary to measure the plasma parameters to characterize a homemade APPJ in interaction with liquid.

Optical emission spectroscopy (OES) [17], as an affordable and a non-intrusive diagnostics method with an easy experimental setup, is used to identify the plasma parameters and reactive species produced by an APPJ. In this case, the radiations emitted by excited atoms, molecules, and reactive species in the plasma source are collected and analyzed to determine the plasma parameters, such as electron density, electron temperature, gas temperature, and so on. The Boltzmann plot [18] is an established method to estimate the electron temperature by assuming that the plasma condition is in a state of partial local thermodynamic equilibrium due to high collision frequency between particles in atmospheric pressure [19]. The upper levels of the atomic transitions follow the Saha–Boltzmann distribution. Hence, the excitation and electron temperatures are assumed to be the same, although this is an inappropriate assumption in mid and low pressures [20,21]. Stark broadening of the Balmer series lines of atomic hydrogen (H_α, H_β, H_γ) [18], which are broadenings or shifts of the spectral lines due to the presence of the electric fields of charged particles, allows us to calculate the electron density in atmospheric pressure. Electron density is almost independent of electron temperature for a given broadening of the H_β line, while in the cases of H_α and H_γ, this dependency is obvious specifically for H_α [22]. Therefore, the H_β broadening (486.1 nm) is utilized to diagnose the electron density. The Stark broadening can be considered as a reliable method for the values of electron density higher than 10^{19} m^{-3}. Otherwise, less than this value, the contributions of Doppler and Van der Waals would be dominant, leading to a high error range in the estimation of electron density [23].

In addition to measurements of electron density and electron temperature, the neutral gas temperature plays a vital role in plasma characterization and processes. For instance, in biological applications, the high gas temperature of the APPJ is capable of damaging biological cells and tissues. Generally, there is no specific threshold temperature for heating damage, which depends on the type of biological cells and tissues; however, in biological testing, the temperature should be less than 42 °C [24]. Therefore, to confirm whether the APPJ might be suitable for the desired application, measurements of the gas temperature would be necessary. Also, to estimate the electron density by the Stark effect, the contributions of Van der Waals and Doppler broadenings, which depend on the gas temperature, should be considered. Hence, knowing the gas temperature is needed. In atmospheric pressure, it is assumed that the rotational temperature of the second positive system of nitrogen gas is equal to the neutral gas temperature [25].

In this study, to characterize and optimize the APPJ, the relative intensities of reactive excited species have been measured using OES. The electron temperature has been estimated by the Boltzmann plot alongside the plasma jet. Since the Balmer series line of atomic hydrogen has been too weak, the measurements of Stark broadening have been carried out at the bottom of the plasma jet. While the flow rate of helium gas is held constant, the influences of argon gas with various flow rates on relative intensities of species and plasma parameters in the presence and the absence of de-ionized (DI) water as a target have been investigated upstream, midstream, and downstream of the plasma jet. The rotational temperature of the second positive system of nitrogen gas has been measured to estimate the neutral gas temperature downstream of the plasma jet in direct interaction with DI water.

2. Materials and Methods

2.1. Experimental Setup of the Atmospheric Pressure Plasma Jet

The experimental setup of the APPJ is shown in Figure 1. The setup consists of a quartz tube with two copper electrodes in a cylindrical shape with a gap distance of 15 mm. A high voltage DC pulsed power supply delivers a monopolar pulsed voltage with a square wave-form pulse. The on-time and off-time of the pulsed voltage were adjusted to be 25 µs, while the rise and fall time was adjusted to be 3 µs. A voltage probe (Rigol-RP1018H) and a current probe (Cybertek-CP8030B) were used to measure the applied voltage and the plasma current. The voltage and current waveform were recorded using an oscilloscope (Rigol DS1054z, 50 MHz, 1 GS/s). During the experiment, helium as a working gas with the flow rate of 5 slm (standard liter/min) remained unchanged, while the argon flow rate as a variable parameter of the working gas was controlled and adjusted to be 0–2000 sccm (standard cubic centimeter/min) by mass flow controller (MFC). The experiments based on the target situation were carried out in three cases. The first case represents the APPJ without de-ionized (DI) water. The second and third cases refer to the presence of DI water as a target exposed by the APPJ directly and indirectly. The maximum distance between the downstream position of the APPJ and the surface of DI water in the case of the indirect plasma exposure was selected to be 4 mm. The APPJ was mounted on an adjustable stage and, thus, the distance of 4 mm was adjusted by the stage.

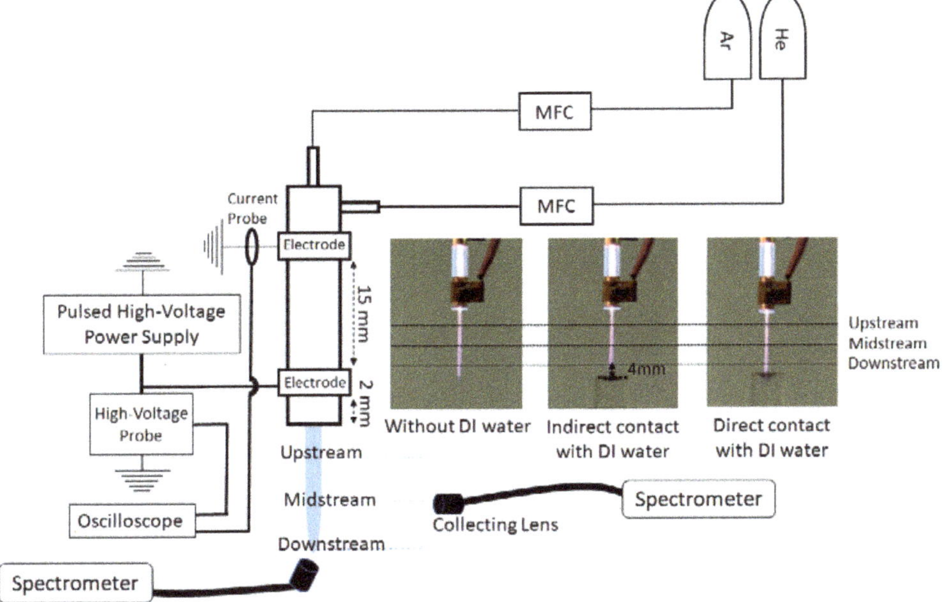

Figure 1. Experimental setup of the helium APPJ with various flow rates of argon gas. The spectroscopic characterization was performed alongside the plasma jet in direct and indirect interaction with and without DI water.

2.2. Electrical Measurements

Figure 2 shows the voltage and current characteristics of the discharge. The maximum values of applied pulsed voltage and frequency were adjusted to be 8.5 kV and 17.8 kHz, respectively. The flow rate of argon gas of the recorded voltage and current was 1600 sccm. It should be noted that the voltage–current waveform was examined with different flow rates of argon gas. The waveform of the discharge did not change significantly.

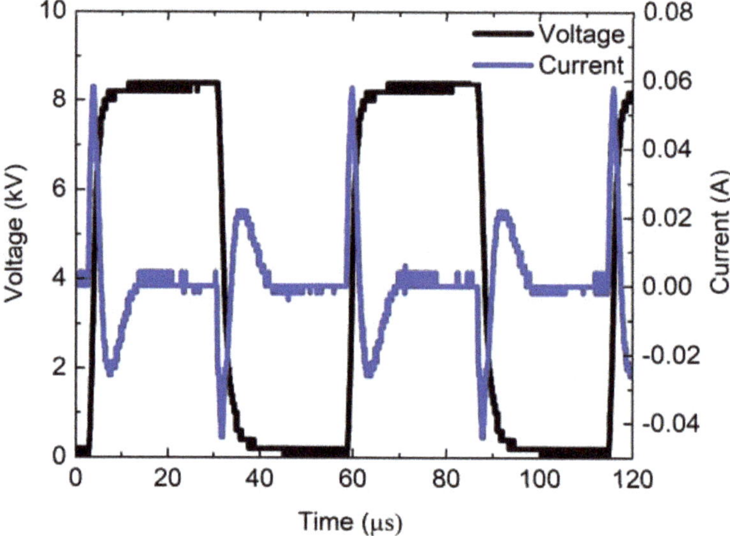

Figure 2. Voltage–current characteristic of the helium-based APPJ with the flow rate of 1600 sccm of argon gas.

2.3. Spectroscopic Measurements

The emission spectra of the APPJ was measured using a spectrometer (Avantes, AvaSpec-2048L, focal length of 75 mm, grating with line density of 300 mm^{-1}, entrance slit of 25 µm, 2048-pixel CCD detector) with a spectral range of 200–1100 nm and a spectral resolution of 1.4 nm. A fiber optics cable including a collecting lens was used to capture the light emitted from the APPJ. The emission spectra were recorded for 5 accumulations with an exposure time of 100 ms. The measuring points of the electron temperature were selected to be upstream, midstream, and downstream of the APPJ as shown in Figure 1. Since the hydrogen Balmer line in the wavelength of 486.1 nm (H$_\beta$) was too weak to be detected in the plasma jet, the spectrometer was placed at the bottom of the plasma jet to enhance the signal intensity of Stark broadening related to the hydrogen Balmer line for measuring the electron density.

3. Results and Discussion

3.1. Measurements of Relative Intensities of Species

Figure 3 presents the spectra of optical emission of the plasma jet obtained by OES in the cases of non-contact (free of DI water—Figure 3a,b), indirect contact (Figure 3c,d), and direct contact (Figure 3e,f) with DI water. The spectra were measured with and without argon feed gas (flow rate of 1600 sccm) downstream and upstream of the jet stream. Hydroxyl (309 nm), helium (706 nm), and argon (750 nm) species, as well as the second positive system of nitrogen gas, were identified in the spectra. As mentioned before and shown in the figure, in all cases, the hydrogen Balmer line (H$_\beta$) is too weak to estimate the electron density.

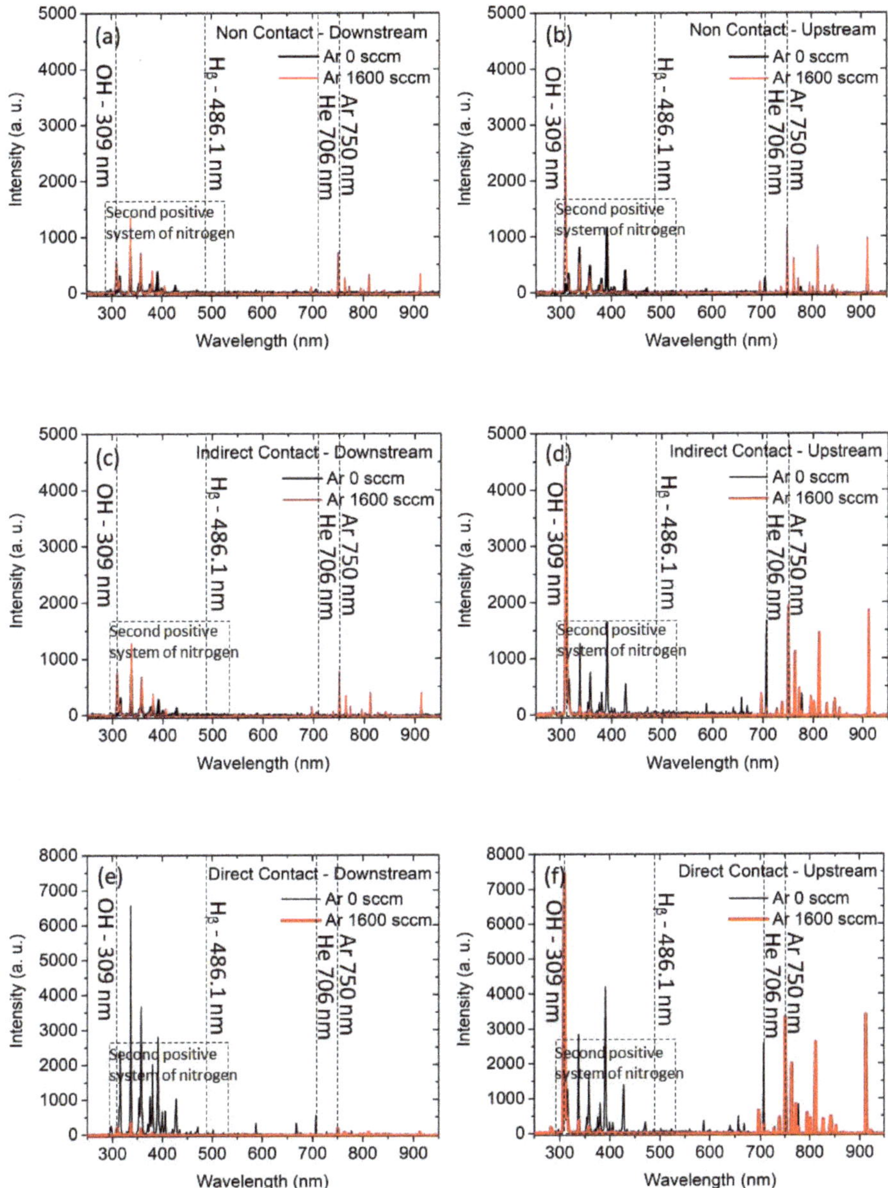

Figure 3. The emission spectra of the plasma jet with and without argon gas in the cases of (**a**) and (**b**) non-contact, (**c**) and (**d**) indirect contact with DI water, and (**e**) and (**f**) direct contact with DI water. The measurements have been performed at the downstream and the upstream of the plasma jet.

Figure 4 shows the spatial profile of relative intensities of hydroxyl (309 nm), helium (706 nm), and argon (750 nm) species of the APPJ obtained by OES with different flow rates of argon gas in the cases of free of DI water (Figure 4a), and indirect contact (Figure 4b) and direct contact with DI water (Figure 4c). The figure shows that relative intensities of hydroxyl are higher upstream of the jet compared to the other measured positions for all cases. In the case of the indirect contact with DI water, shown in Figure 4b, the relative intensity of hydroxyl increases more. The enhancement of the hydroxyl concentration is boosted in the case of direct contact with DI water alongside the plasma jet. The intensity value of hydroxyl species rises more than two times compared to the case of free of DI water by considering the values shown in Figure 4a,c. Moreover, the relative intensities of hydroxyl species reach the maximum amount at the flow rate of 1600 sccm specifically at the upstream of the plasma jet; then, the intensities of hydroxyl decrease for all cases. In the case of direct contact, at the flow rate of 2000 sccm, the spectrum was not available to measure the intensities of the species as the length of the plasma jet decreased to half. Since the flow rate of helium gas was held constant during the experiment, the intensities of helium species remained nearly unchanged, as shown in Figure 4, for all cases.

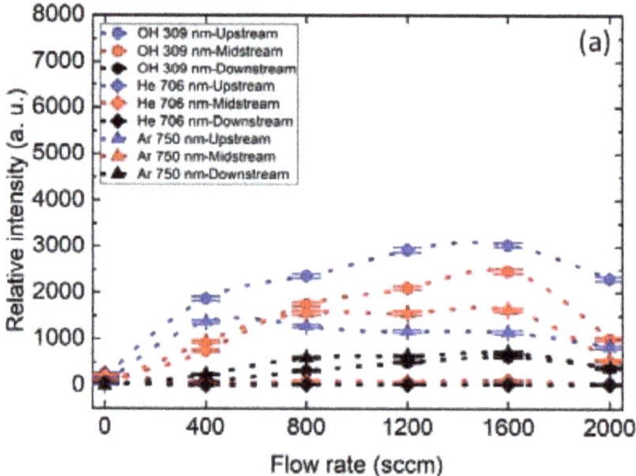

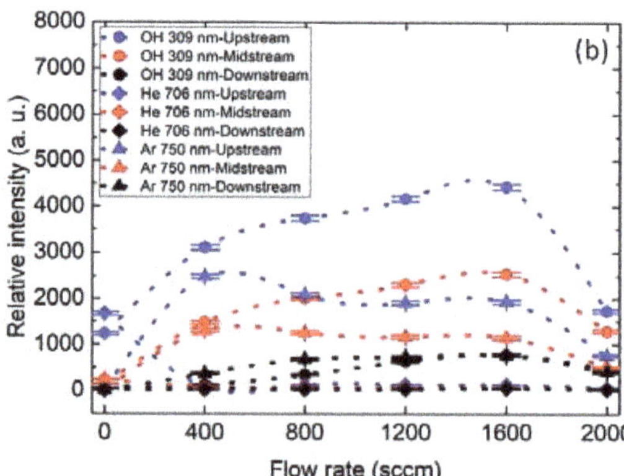

Figure 4. *Cont.*

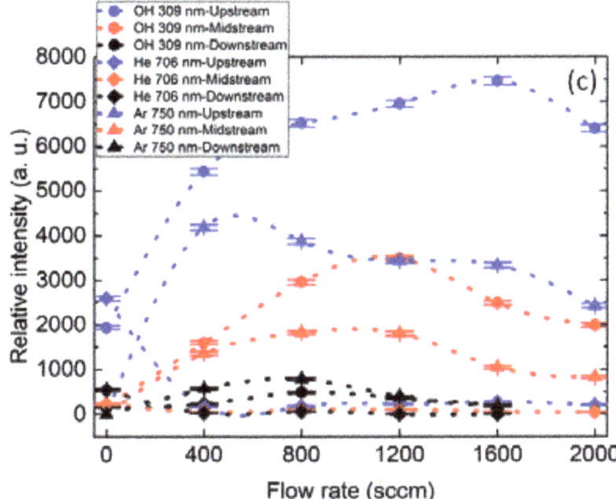

Figure 4. The relative intensity of hydroxyl (309 nm), helium (706 nm), and argon (750 nm) obtained by OES as a function of the flow rate of argon gas in the cases of (**a**) non-contact with DI water; (**b**) indirect contact with DI water; and (**c**) direct contact of DI water.

3.2. Measurements of Gas Temperature

Although the second positive system of nitrogen bands were obvious in all cases (Figure 3), the intensity of the second positive system was strong without argon gas feed in the case of direct contact with DI water as shown in Figure 3e,f. As mentioned above, the nitrogen band allows us to measure the temperature of the neutral particle based on the rotational temperature of nitrogen gas. By fitting the emission line spectra of the second positive system (C $^3\Pi_u \rightarrow$ B $^3\Pi_g$ = 375–381 nm) [26], the gas temperature was measured to be 0.03 eV, equal to 75 °C. Thus, the APPJ is in the non-thermal state and regarded as a cold plasma. In addition, the ambient temperature was measured by a mercury thermometer at a distance of around 10 mm from the tip of the plasma jet for 2 minutes. The temperature was recorded to be 33.2 °C.

3.3. Measurements of Electron Density and Electron Temperature

Figure 5 illustrates the spatial profile of the electron temperature estimated by the Boltzmann plot upstream, midstream, and downstream of the APPJ in the cases of free of DI water (Figure 5a), and indirect (Figure 5b) and direct contact with DI water (Figure 5c). The measured parameters are plotted with different flow rates of argon gas. The estimated electron density based on Stark broadening in the case of free of DI water (no target) is shown in Figure 5a. At the flow rate of 400 sccm, the electron density and the electron temperature were estimated to be 1.2×10^{22} m^{-3} and 0.25–0.45 eV alongside the plasma jet. The electron density reached the maximum value of 2.3×10^{22} m^{-3} at the flow rate of 1600 sccm, then decreased to the flow rate of 2000 sccm. However, the trend of electron temperature was downward between the flow rates of 400 to 2000 sccm, and finally, reached the values of 0.15–0.21 eV at the flow rate of 2000 sccm. Injecting the massive species to the plasma discharge leads to a change in the energy transfer between electrons and the species due to the collision frequency that causes a reduction in electron temperature. The electron temperature converges more by increasing the flow rate of argon gas in the case of direct contact; nevertheless, the trend of electron temperature is almost the same for all cases. In the case of direct contact, at the flow rate of 2000 sccm, the spectrum was not available to estimate the electron temperature as the length of the plasma jet decreased to half.

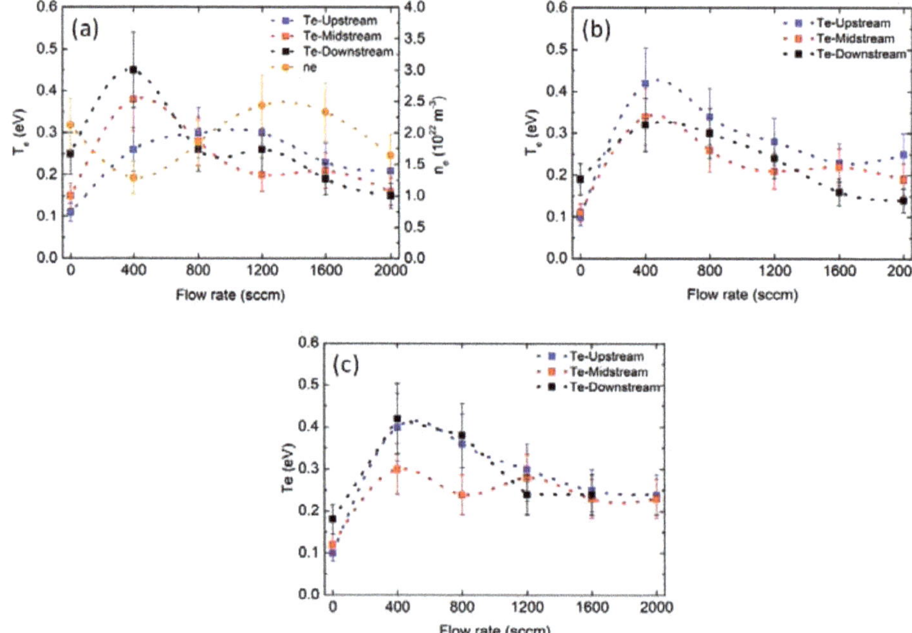

Figure 5. Spatial profile measurements of the electron temperature upstream, midstream, and downstream of the APPJ in the cases of (**a**) non-contact with DI water; (**b**) indirect contact with DI water; and (**c**) direct contact of DI water. Measurement of electron density was undertaken at the bottom of the jet stream.

The Boltzmann plot equation is assumed to be $\ln(I\lambda/gA) = -E/kT_e$ + constant, where I is the measured intensity obtained by OES, λ is the selected wavelength associated with helium and argon gases, g is the statistical weight, A is the transition probability, E is the excitation energy corresponding to the selected wavelengths, k is Boltzmann constant, and T_e is electron temperature. The selected wavelengths of atomic emission lines related to argon and helium gases were obtained from references [12,27] to estimate the electron temperature based on the Boltzmann plot. The spectroscopic data related to the wavelengths of excited atoms and ions were confirmed via the NIST atomic database [28]. By adding argon gas, the contributed wavelengths of helium [27] for the estimation of electron temperature weakened; therefore, the wavelengths of argon [12] were used at the flow rate of 400 to 2000 sccm.

Regarding Stark broadening, although hydrogen gas was not used as a working gas in the APPJ, it initially might result from the impurity of helium and argon working gases or the ambient air. The relationship between electron density and Stark broadening is assumed to be $n_e = (\Delta\lambda_{\mathrm{FWHM}}/(2 \times 10^{-11}))^{3/2}$, where $\Delta\lambda_{\mathrm{FWHM}}$ is Stark broadening and n_e is electron density in terms of per cubic centimeter. The broadening of the spectral line is a convolution of instrumental, Doppler, and Van der Waals broadenings [29]. The Voigt fitting of the measured Stark bordering was performed using Origin Pro 5. The instrumental broadening was obtained by replacing the APPJ with a mercury lamp (wavelength of 546.1 nm) at low pressure. This value was obtained to be 0.06 nm. Finally, the Voigt fitting was excluded from the broadenings mentioned above to estimate the pure Stark broadening at full width at half maximum (FWHM) for measurements of electron density.

The presence of massive particles such as argon species in the helium plasma changes the plasma chemistry and the degree of ionization. Based on the results of relative intensities of excited species obtained by OES, the hydroxyl concentration is highest upstream of the plasma jet compared to the other regions. The production of hydroxyl might be explained by two possible different mechanisms of reactions [30]. The first mechanism corresponds to the direct dissociation of the water molecule by energetic electrons. Based on the results, by increasing the flow rate, the electron temperature drops. So, the energy of the electrons is too low (less than 1 eV) to dissociate the water molecule. The second refers to the possibility of participation of metastable argon to produce H_2O^+. In this case, the dissociative recombination of H_2O^+ leads to hydroxyl generation. Therefore, the metastable atoms may play an important role to generate hydroxyls [14,31]. That is the reason the concentration of hydroxyl is highest upstream of the plasma jet for all cases, as the concentration of the metastable species of argon is highest in the same space. In addition, as the plasma jet approaches the water surface, the humidity around the plasma jet enhances due to the water evaporation that causes a boost of hydroxyl generation in the presence of water. Yang et al. [30] reported similar results by using the helium-based APPJ with the fixed flow rate of helium gas, while in our case, the flow rate of the APPJ changed by argon gas.

4. Conclusions

We characterized and optimized a helium APPJ with a mixture of argon gas in the presence and the absence of DI water. DI water as a target was exposed by the APPJ directly and indirectly. Electron temperature and electron density as a function of the flow rate of the gas mixtures were investigated. In a direct interaction of the jet stream and surface of DI water, the gas temperature was estimated based on the rotational temperature to confirm the plasma condition is in the non-thermal regime. The OES results show that approaching the APPJ so that the jet stream would have an indirect interaction with the water surface leads to more hydroxyl concentration with respect to the free plasma case. Hydroxyl production could be boosted more in direct contact with DI water. Considering positive points and drawbacks of using pure helium or argon gases to produce the APPJ, mixing these two gases might be the right solution for biological and environmental applications.

Author Contributions: Formal analysis, N.B.; Investigation, Y.-Z.Y.; Supervision, J.-H.H. and C.L.; Writing–original draft, N.B.; Writing–review & editing, N.B.

Funding: This research received no external funding.

Conflicts of Interest: The authors declare no conflict of interest.

References

1. Winter, J.; Brandenburg, R.; Weltmann, K.-D. Atmospheric pressure plasma jets: An overview of devices and new directions. *Plasma Source Sci. Technol.* **2015**, *24*, 064001. [CrossRef]
2. Pedroni, M.; Morandi, S.; Silvetti, T.; Cremona, A.; Gittini, G.; Nardone, A.; Pallotta, F.; Brasca, M.; Vassallo, E. Bacteria inactivation by atmospheric pressure plasma jet treatment. *J. Vac. Sci. Technol.* **2017**, *36*, 01A107. [CrossRef]
3. Gilmore, B.F.; Flynn, P.B.; Brien, S.O.; Hickok, N.; Freeman, T.; Bourke, P. Cold plasmas for biofilm control: Opportunities and challenges. *Trends Biotechnol.* **2018**, *36*, 627–628. [CrossRef] [PubMed]
4. Keidar, M.; Shashurin, A.; Volotskova, O.; Stepp, M.A.; Srinivasan, P.; Sandler, A.; Trink, B. Cold atmospheric plasma in cancer therapy. *Phys. Plasmas* **2013**, *20*, 057101. [CrossRef]
5. Foster, J.E. Plasma-based water purification: Challenges and prospects for the future. *Phys. Plasmas* **2017**, *24*, 055501. [CrossRef]
6. Thirumdas, R.; Kothakota, A.; Annapure, U.; Siliveru, K.; Blundell, R.; Gatt, R.; Valdramidis, V.P. Plasma activated water (PAW): Chemistry, physio-chemical properties, applications in food and agriculture. *Trends Food Sci. Technol.* **2018**, *77*, 21–31. [CrossRef]
7. Laroussi, M.; Lu, X. Room-temperature atmospheric pressure plasma plume for biomedical applications. *Appl. Phys. Lett.* **2005**, *87*, 113902. [CrossRef]
8. Bruggeman, P.; Brandenburg, R. Atmospheric pressure discharge filaments and microplasmas: Physics, chemistry and diagnostics. *J. Phys. D Appl. Phys.* **2013**, *46*, 464001. [CrossRef]

9. Santosh, V.S.; Kondetia, K.; Phanb, C.Q.; Wendec, K.; Jablonowskic, H.; Gangala, U.; Granickd, J.L.; Hunterb, R.C.; Bruggemana, P.J. Long-lived and short-lived reactive species produced by a cold atmospheric pressure plasma jet for the inactivation of Pseudomonas aeruginosa and Staphylococcus aureus. *Free Radic. Biol. Med.* **2018**, *124*, 275–287. [CrossRef]

10. Liu, C.T.; Wu, C.J.; Yang, Y.W.; Lin, Z.H.; Wu, J.S.; Hsiao, S.C.; Lin, C.P. Atomic Oxygen and Hydroxyl Radical Generation in Round Helium-Based Atmospheric-Pressure Plasma Jets by Various Electrode Arrangements and Its Application in Sterilizing Streptococcus mutans. *IEEE Trans. Plasma Sci.* **2014**, *42*, 12. [CrossRef]

11. Yue, Y.F.; Mohades, S.; Laroussi, M.; Lu, X. Measurements of Plasma-Generated Hydroxyl and Hydrogen Peroxide Concentrations for Plasma Medicine Applications. *IEEE Trans. Plasma Sci.* **2016**, *44*, 11. [CrossRef]

12. Sarani, A.; Nikiforov, A.Y.; Leys, C. Atmospheric pressure plasma jet in Ar and Ar/H2O mixtures: Optical emission spectroscopy and temperature measurements. *Phys. Plasmas* **2010**, *17*, 063504. [CrossRef]

13. Cheng, C.; Shen, J.; Xiao, D.-Z.; Xie, H.-B.; Lan, Y.; Fang, S.-D.; Meng, Y.-D.; Chu, P.K. Atmospheric pressure plasma jet utilizing Ar and Ar/H2O mixtures and its applications to bacteria inactivation. *Chin. Phys. B* **2014**, *23*, 0752204. [CrossRef]

14. Nikiforov, A.Y.; Sarani, A.; Leys, C. The influence of water vapor content on electrical and spectral properties of an atmospheric pressure plasma jet. *Plasma Source Sci. Technol.* **2011**, *20*, 015014. [CrossRef]

15. Wang, S.; Schulz-von der Gathen, V.; Döbele, H.F. Discharge comparison of nonequilibrium atmospheric pressure Ar/O2 and He/O2 plasma jets. *Appl. Phys. Lett.* **2003**, *83*, 16. [CrossRef]

16. Li, S.-Z.; Lim, J.-P.; Kang, J.G.; Uhm, H.S. Comparison of atmospheric-pressure helium and argon plasmas generated by capacitively coupled radio-frequency discharge. *Phys. Plasmas* **2006**, *13*, 093503. [CrossRef]

17. Fantz, U. Basic of plasma spectroscopy. *Plasma Source Sci. Technol.* **2006**, *15*, S137–S147. [CrossRef]

18. Griem, H.R. *Plasma Spectroscopy*; McGraw-Hill: New York, NY, USA, 1964.

19. Calzada, M.D.; Moisan, M.; Gamero, A.; Sola, A. Experimental investigation and characterization of the departure from local thermodynamic equilibrium along a surface-wave-sustained discharge at atmospheric pressure. *J. Appl. Phys.* **1996**, *80*, 46. [CrossRef]

20. Sola, A.; Calzada, M.D.; Gamero, A. On the use of the line-to-continuum intensity ratio for determining the electron temperature in a high-pressure argon surface-microwave discharge. *J. Phys. D Appl. Phys.* **1995**, *28*, 4. [CrossRef]

21. Gordillo-Vázquez, F.J.; Camero, M.; Gómez-Aleixandre, C. Spectroscopic measurements of the electron temperature in low pressure radiofrequency Ar/H2/C2H2 and Ar/H2/CH4 plasmas used for the synthesis of nanocarbon structures. *Plasma Sources Sci. Technol.* **2005**, *15*, 1. [CrossRef]

22. Torres, J.; Palomares, J.M.; Sola, A.; van der Mullen, J.J.A.M.; Gamero, A. A Stark broadening method to determine simultaneously the electron temperature and density in high-pressure microwave plasmas. *J. Phys. D Appl. Phys.* **2007**, *40*, 5929–5936. [CrossRef]

23. Zhu, X.M.; Pu, Y.K.; Balcon, N.; Boswell, R. Measurement of the electron density in atmospheric-pressure low-temperature argon discharges by line-ratio method of optical emission spectroscopy. *J. Phys. D Appl. Phys.* **2009**, *42*, 142003. [CrossRef]

24. Yarmolenko, P.S.; Moon, E.J.; Landon, C.; Manzoor, A.; Hochman, D.W.; Viglianti, B.L.; Dewhirst, M.W. Thresholds for thermal damage to normal tissues: An update. *Int. J. Hyperth.* **2011**, *27*, 320–343. [CrossRef] [PubMed]

25. Masoud, N.; Martus, K.; Figus, M.; Becker, K. Rotational and Vibrational Temperature Measurements in a High-Pressure Cylindrical Dielectric Barrier Discharge (C-DBD). *Contrib. Plasma Phys.* **2005**, *45*, 30–37. [CrossRef]

26. Koike, S.; Sakamoto, T.; Kobori, H.; Matsuura, H.; Akatsuka, H. Spectroscopic Study on Vibrational Nonequilibrium of a Microwave Discharge Nitrogen Plasma. *Jpn. J. Appl. Phys.* **2004**, *43*, 5550. [CrossRef]

27. Gulec, A.; Bozduman, F.; Hala, A.M. Atmospheric pressure 2.45-GHz microwave helium plasma. *IEEE Trans. Plasma Sci.* **2015**, *43*, 786790. [CrossRef]

28. NIST Atomic Spectra Database. Available online: http://physics.nist.gov/PhysRefData (accessed on 21 February 2019).

29. Ouyang, Z.; Surla, V.; Cho, S.T.; Ruzic, D.N. Characterization of an atmospheric-pressure helium plasma generated by 2.45-GHz microwave power. *IEEE Trans. Plasma Sci.* **2012**, *40*, 3476–3481. [CrossRef]

30. Yang, Y.; Zhang, Y.; Liao, Z.; Pei, X.; Wu, S. OH Radicals Distribution and Discharge Dynamics of an Atmospheric Pressure Plasma Jet above Water Surface. *IEEE Trans. Radiat. Plasma Med. Sci.* **2018**, *2*, 223–228. [CrossRef]

31. Collette, A.; Dufour, T.; Reniers, F. Reactivity of water vapor in an atmospheric argon flowing post discharge plasma torch. *Plasma Sources Sci. Technol.* **2016**, *25*, 025014. [CrossRef]

Article

Experimental Investigation on the Influence of Target Physical Properties on an Impinging Plasma Jet

Emanuele Simoncelli [1], Augusto Stancampiano [1,2], Marco Boselli [1], Matteo Gherardi [1,3] and Vittorio Colombo [1,3,*]

1   Department of Industrial Engineering (DIN), Alma Mater Studiorum-Università di Bologna, 40131 Bologna, Italy; emanuele.simoncelli@unibo.it (E.S.); augusto.stancampiano@univ-orleans.fr (A.S.); marco.boselli@unibo.it (M.B.); matteo.gherardi4@unibo.it (M.G.)
2   Presently at GREMI, UMR7344 CNRS/Université d'Orléans, 45067 Orléans, France
3   CIRI—Advanced Applications in Mechanical Engineering & Materials Technology, Alma Mater Studiorum—Università di Bologna, 40131 Bologna, Italy
*   Correspondence: vittorio.colombo@unibo.it; Tel.: +39-051-209-3978

Received: 15 July 2019; Accepted: 11 September 2019; Published: 16 September 2019

Abstract: The present work aims to investigate the interaction between a plasma jet and targets with different physical properties. Electrical, morphological and fluid-dynamic characterizations were performed on a plasma jet impinging on metal, dielectric and liquid substrates by means of Intensified Charge-Coupled Device (ICCD) and high-speed Schlieren imaging techniques. The results highlight how the light emission of the discharge, its time behavior and morphology, and the plasma-induced turbulence in the flow are affected by the nature of the target. Surprisingly, the liquid target induces the formation of turbulent fronts in the gas flow similar to the metal target, although the dissipated power in the former case is lower than in the latter. On the other hand, the propagation velocity of the turbulent front is independent of the target nature and it is affected only by the working gas flow rate.

Keywords: impinging jet; metal/dielectric/liquid substrate; ICCD imaging; high-speed Schlieren imaging; plasma discharge morphology; turbulence

1. Introduction

The physical and chemical properties of a cold atmospheric pressure plasma (CAP) jet are not uniquely dependent on the plasma source configuration and operational parameters, but also on the target characteristics. The complex mutual interaction between the plasma and the target has recently been the subject of an increasing number of papers, investigating how the targets significantly affect the plasma properties, such as fluid-dynamics [1–3], electromagnetic field [4,5], reactive and excited species production and distribution [6], and ionization front velocity and propagation [1]. Especially, electrical properties, such as conductivity and potential, play a major role [4,7]. During direct plasma treatment, the target becomes part of the transient electrical circuit, connecting the power supply, the plasma source, the plasma, the target and the return to ground [8,9]. This means that the target's electrical parameters (conductivity, potential, etc.) play a major role in determining treatment conditions. Darny et al. revealed how, with conductive substrates like metal, the electric field adjacent to the substrate is enhanced, allowing for a restrike discharge that in turn can greatly enhance the production of reactive species that play an important role in biological applications [10]. Modelling studies by Norberg et al. also predicted similar modifications of the electric field based on substrate conductivity [11]. As reported by Yuanfu et al. [7], the production of OH radicals over a metal substrate can be ten times higher than in the case of a dielectric substrate. In light of the wide range of industrial, biomedical and agricultural applications, the interaction of atmospheric plasma jets with liquids can be of crucial interest to the scientific community. Liquid substrates, having both capacitive and conductive components, show a hybrid behavior between the purely dielectric and the purely conductive materials [1,7]. This behavior is modulated by the conductivity of liquid solutions that can greatly vary in a range of several order of magnitude (10^{-7}–10^{-2} S/cm).

In this frame, the present work provides a direct comparison of the behavior of an atmospheric pressure plasma jet when interacting with a conductive liquid solution, or with dielectric and metal substrates, using ICCD and high-speed Schlieren as imaging techniques. The aim is to gather new insight into the fluid-dynamic behavior of this configuration; an aspect which has not been deeply investigated in the literature, despite being of great influence in plasma assisted processes.

2. Materials and Methods

2.1. Plasma Source

The plasma source adopted in this work was a single electrode plasma jet developed at the University of Bologna, Italy, and already described, characterized, and applied in previous works by Colombo et al. [1,2,12]. The plasma source was driven by a commercial nanosecond pulse generator (FPG 20-1NMK, FID GmbH). The electric conditions used for all experiments were 15 kV as the peak voltage (PV) and 125Hz as the pulse repetition frequency (PRF). The main voltage pulse supplied by the generator lasted around 35 ns with 10 ns as rise time. A residual damping signal was recorded up to 100 ns, thus the discharge event lasted less than 100 ns. The ignition of the discharge was repeated every 8 ms as a consequence of imposed PRF.

The high-voltage pin electrode (a stainless steel needle; Ø 0.3 mm) was centered inside a dielectric channel, and a flow rate of 3 slpm of helium gas (99.999% pure) was injected through a 12 hole (Ø 0.3 mm) diffuser. The plasma was ejected from the source into the surrounding atmosphere through an orifice with a diameter of 1 mm, producing a visible plasma plume and possibly interacting with a substrate. All experiments were performed in controlled ambient air at 30 °C. To estimate the electrical power dissipated by the plasma source, voltage and current waveforms were recorded on the high-voltage connection powering the plasma source by means of a high-voltage probe (Tektronix P6015A, sensitivity of 0.018%) and a current probe (Pearson 6585, sensitivity of 1%) connected to an oscilloscope (Tektronix DPO40034). The power density was evaluated using the following formula:

$$SPD = PRF \cdot \int V \cdot I dt \tag{1}$$

where *PRF* is the pulse repetition frequency, *V* is the applied voltage, and *I* is the current (the voltage and current signal are related to the main positive peak).

2.2. Substrates

The plasma source was positioned vertically at 10 mm (fixed gap) above the substrate surface. Different targets characterized by different conductivities were selected for this study, ranging from metal and liquid to dielectric substrates. A stainless steel plate (7.6 cm × 7.6 cm × 1 cm) was chosen as the target with almost infinite conductivity (~3.7×10^{11} µS/cm). On the other hand, to simulate a non-conductive substrate, a 7.6 cm × 7.6 cm × 1 cm PVC plate was used as the dielectric substrate (~10^{-17} µS/cm). As far as the liquid substrate was concerned, since the electrical conductivity of the target affects the plasma characteristics [13,14], and in turn the plasma treatment may alter the electrical conductivity of the treated liquid solutions [15,16], the liquid target was prepared as a phosphate buffer solution (made by dissolving sodium phosphate dibasic (Na_2HPO_4) and potassium phosphate monobasic (KH_2PO_4) in distilled water). A solution characterized by an electrical conductivity of 119 µS/cm with a pH of 7.2 was realized. The physical properties (conductivity and pH) were monitored and remained unaltered during the experiments. The solution volume was 120 mL contained in a vessel (7.6 cm × 7.6 cm × 2 cm) with quartz sidewalls and an aluminum bottom. Since the metal substrate and the aluminum bottom of the liquid substrate vessel were connected to ground through a low impedance electrical connection, both could be considered at ground potential. The dielectric substrate was instead positioned on a grounded metal plate to control and fix the associated capacitance. According to the literature [9,17], the equivalent electrical circuits associated with the three substrates result in a single resistance for the metal substrate, a single capacitance for the dielectric substrate, and a resistance and capacitance in parallel for the liquid substrate. Presenting very different electrical characteristics, the three substrates are a good representation of a wide range of possible targets.

2.3. Diagnostic Techniques

The morphology and the time evolution of the plasma discharge interacting with each substrate were investigated by means of an ICCD camera (Princeton Instruments PIMAX3, spectral response 180–900 nm) equipped with a conventional macro lens (Sigma Dg-Ex-APO-If 180 Mm/F3.5, spectral response 380–900 nm). To synchronize the image acquisition with the discharge event, the synchronization of the camera gating, driven by the voltage pulse, was performed by employing a delay generator (BNC 575 digital pulse/delay generator) and taking into account all possible signal transmission delays, as already described in the literature [12]. An overview of the iCCD imaging setup is showed in Figure 1.

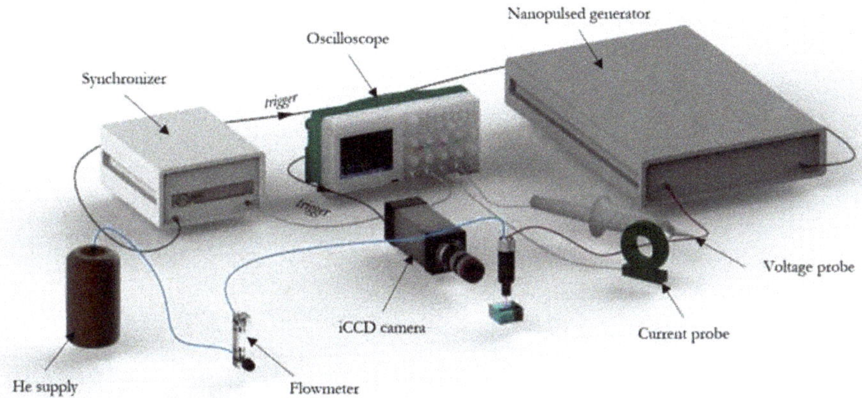

Figure 1. ICCD imaging setup.

The overall emission intensity, produced by the plasma discharge during the main voltage pulse, was acquired by imposing both the camera gate exposure (35 ns) and the duration of the voltage pulse. The images were accumulated 30 times with a gain factor set at 50. The time evolution of the plasma discharge during the whole voltage pulse was investigated through the capture of sequential 10 ns camera gate exposures (each frame resulting from 30 accumulations). The first ICCD gate opening (0 ns) was imposed, so as to center on the start of the rising front of the voltage pulse (see Figure 3, top). The ICCD camera gates were superimposed on the excitation voltage waveforms. The reproducibility of the discharge was first verified with single shot (no accumulations) acquisitions (data not shown).

The characterization of the fluid-dynamic of the impinging jet was performed using the Schlieren imaging technique [18]. The Schlieren setup is shown in Figure 2 and was composed of a 450 W ozone free xenon lamp (Newport-Oriel 66355 Simplicity Arc Source) as a light source, a slit and an iris diaphragm, two parabolic mirrors with a focal length of 1 m, a knife edge positioned vertically because the highest gradients of the refractive index around the axis of the jet were horizontal, and a high-speed camera (Memrecam GX-3 NAC image technology). The camera was operated at 8000 fps with 1/200,000 s shutter time. The plasma source was positioned halfway between the two parabolic mirrors.

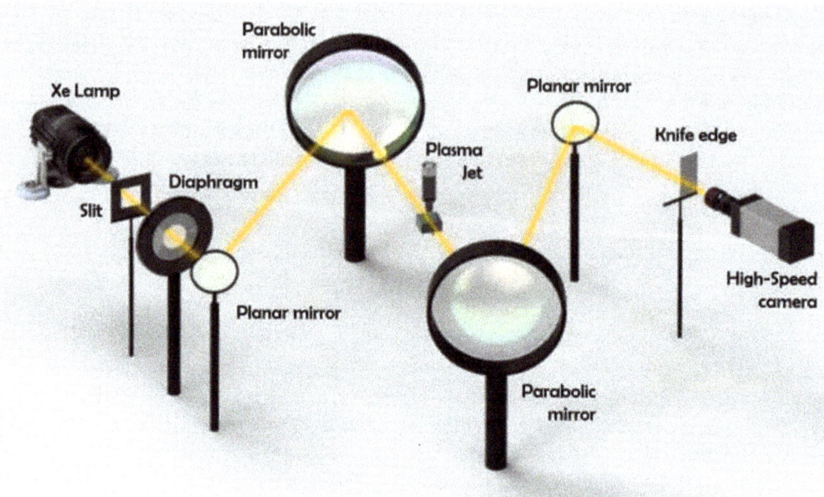

Figure 2. Schlieren imaging setup.

3. Results

3.1. Electrical and Time-Resolved ICCD Characterization of the Plasma Jet Impinging on Different Substrates

Table 1 shows the measured values of the electrical power, dissipated by the plasma source interacting with metal, dielectric and liquid substrates, respectively. Although the input operating conditions were the same for all investigated cases, the plasma jet impinging on a metal target dissipated the highest power (0.434 W) in comparison to the other targets.

Table 1. Electrical power dissipated by the plasma source for metal, dielectric and liquid substrates.

	Metal	**Dielectric**	**Liquid (119 µS/cm)**
Power [W]	0.434	0.313	0.255

In this section, the results for the ICCD imaging of the plasma jet plume impinging on different substrates are presented.

Figure 3 shows, for each condition, a single accumulated image of the discharge obtained by capturing the light intensity emitted during the whole main voltage pulse (duration of ~35 ns). The acquisitions highlight how the plasma, and therefore its Visible–Near InfraRed (Vis-NIR) light emission, was highly influenced by the nature of the substrate. The strongest intensity and largest width of the plasma columns were recorded for the case of a metal target (fourth image from the left in Figure 3). On the other hand, the lowest intensity was observed when the plasma jet impinged on the liquid substrate (second image from the left in Figure 3). Furthermore, a spreading of the plasma across the target surface, known as surface ionization wave (SIW), was clearly observed only in the case of the dielectric substrate. For the liquid substrate, the plasma discharge appeared to be focused on a small area, corresponding to the dimple created in the liquid by the jet effluent (first image on the left in Figure 3).

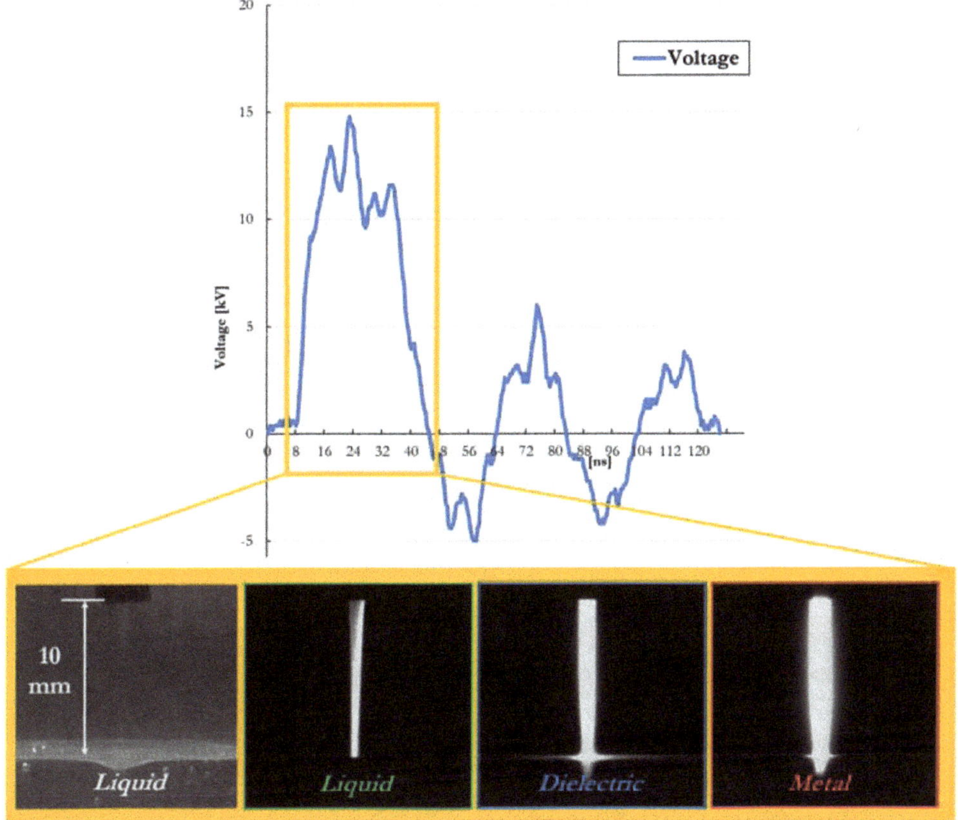

Figure 3. ICCD images of the discharge emission related to the dielectric, liquid and metal substrates. (The first image on the left, taken with a longer exposure time, is presented to show the dimple formation on the liquid substrate).

In Figure 4, the sequences of ICCD images (10 ns exposure) for both dielectric and metal cases are shown.

The acquisitions (Figure 4) highlight how the temporal evolution of the plasma Vis-NIR light emission was influenced by the nature of the target. The propagation velocity of the ionization front was estimated to be $\geq 2 \times 10^8$ cm/s, as the crossing of the gap took place in the first two acquisitions.

For both substrates, the peak of the emission intensity was achieved in the acquisition at 20 ns, corresponding to the reach of the maximum applied voltage (Figure 4, third acquisition from the left for both cases). In the case of the dielectric target, the images show how the Vis-NIR light emission remained approximately constant during the whole voltage pulse. In contrast, on the metal substrate, the plasma emission significantly changed with time; therefore, the higher the applied voltage, the more intense the light emission.

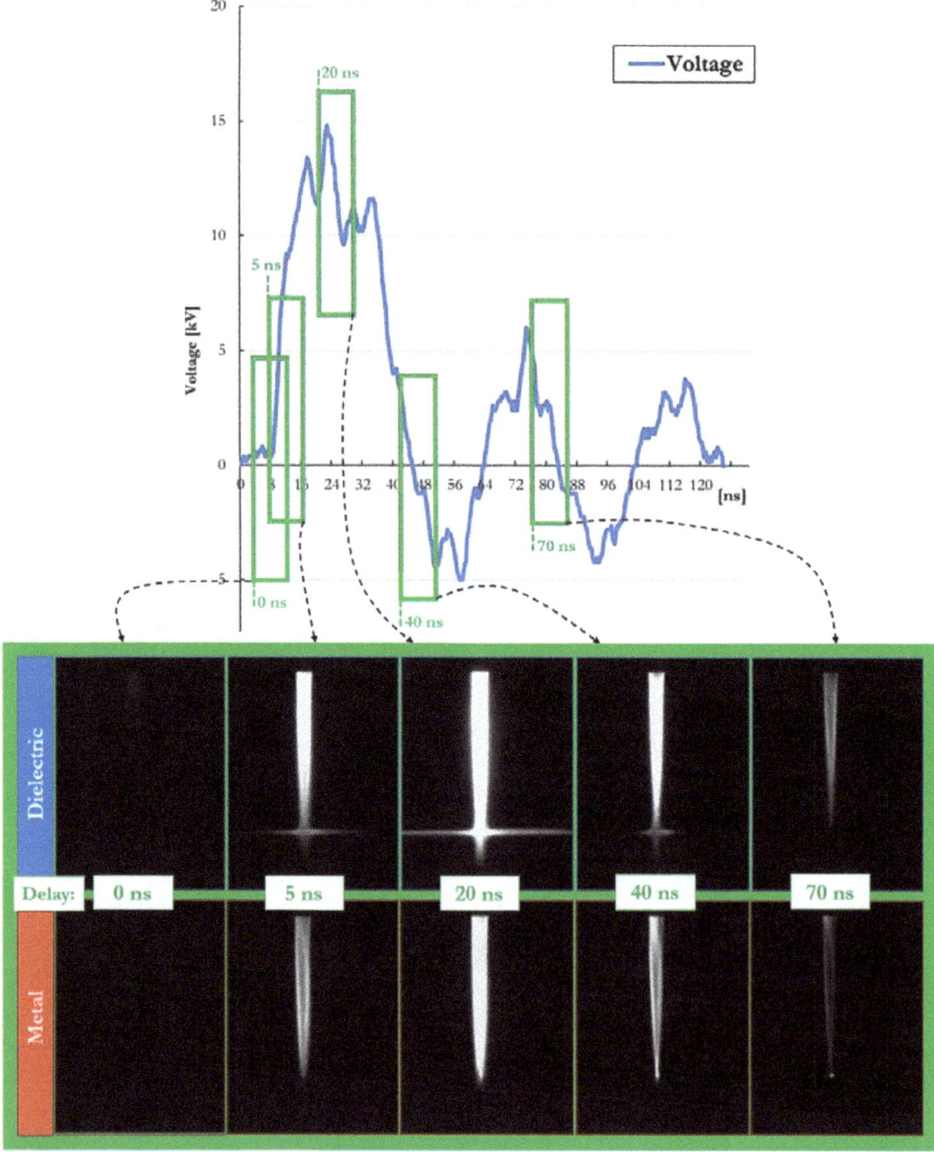

Figure 4. Time-resolved ICCD imaging of the plasma jet plume impinging on the dielectric and metal substrates.

The ICCD images in Figure 4 reveal how, in the case of a metal substrate, the plasma plume did not show a clear SIW formation, but the discharge was quite focused on a point on the surface. On the other hand, in Figure 3 the image related to the metal substrate shows a weak SIW upon the surface, as a consequence of a time integration of 35 ns corresponding to the imposed exposure time. As shown by the frames 5 and 20 ns in Figure 4, (metal substrate) photons emission covered a wider area compared to the frames of 40 and 70 ns.

In the ICCD time-resolved images shown in Figure 4, the SIW formation and development were clearly enhanced in the case of the plasma jet interacting with the dielectric target, due to its higher capacitance. The dielectric surface was charged more than the metal one, favoring a higher spreading of the plume over it.

3.2. Time-Resolved Schlieren Characterization of the Plasma Jet Impinging on Different Substrates

Figure 5 presents results for the high-speed Schlieren imaging of the plasma jet plume impinging on the dielectric, liquid and metal substrates. The frames were selected with the aim of emphasizing the most important steps of spatial evolution of the turbulent front induced by the discharge, as follows: the discharge event, the turbulent front exiting the nozzle, the turbulent front approaching the substrate, its impact with the substrate surface, and its expansion upon the surface.

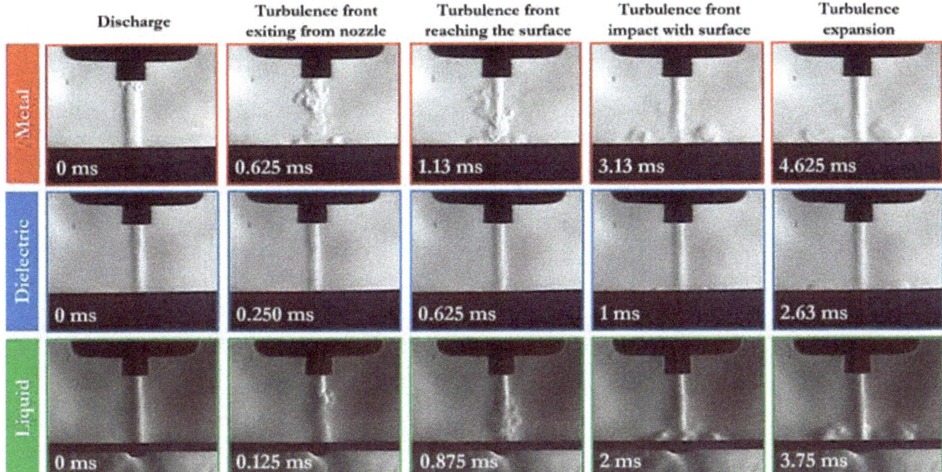

Figure 5. Time-resolved Schlieren images of the plasma jet plume impinging on the metal, dielectric and liquid substrates. The reported time values are indicative of the time lapse from the ns-discharge to the start of the corresponding acquisition of the high-speed camera.

The duration of the plasma discharge emission (around 100 ns, as described in Section 2.1) was entirely captured in the first frame (0 ms), since the exposure of the high-speed camera was set at 0.005 ms. The time values reported on each frame were indicative of the time lapse between the acquisition of the first frame and those that followed.

In the case of the metal target, a turbulent front was observed at the outlet of the plasma jet in the first image, while for all the other cases, in the first frame (0 ms) the He gas flow seemed completely laminar, similar to the case of He gas flow without plasma ignition (data not presented). For dielectric and liquid substrates, the acquisitions at 0 ms showed how during the plasma discharge, the He gas flow was laminar with no flow modifications visible. Nonetheless, a significant perturbation of the He gas flow was clearly visible in the following frames, several tens of microseconds after the plasma discharge and the imposed voltage pulse ended. The turbulent phenomena is thus ignited by the discharge event, and its dynamic behavior is directly affected by the nature of the target.

At 0.25 ms after the end of the plasma discharge, in all investigated cases, a transient turbulent structure appeared as a consequence of an induced alteration of the helium gas characteristics, such as temperature and density, inside the plasma source during the discharge. The turbulent front coming out of the nozzle propagated downstream with a velocity close to that of the gas flow (~60 m/s). Once this transient turbulent structure reached the target surface, eddies were generated and propagated above the surface, away from the column axis. In all investigated cases, a laminar flow was finally re-established (data not shown) in the He column before the next discharge was ignited (8 ms period for PRF 125 Hz).

The strongest turbulent front was observed in the case of the metal substrate. The turbulent structures remained visible more than 5 ms after the discharge event (data not shown). In the case of the liquid substrate, the turbulent phenomenon that was induced was well recognizable. On the other hand, when the plasma jet impinged on a dielectric substrate, the induced turbulence was weaker, and the turbulent structures propagated over the surface were rapidly dissipated by interaction with the surrounding air. While the amplitude and the intensity of the turbulence were dependent on the target nature, the velocity of propagation of the turbulent front appeared independent of the target nature and governed by the gas flow rate only.

Finally, the gas impinging on the liquid substrate caused the formation of an axisymmetric dimple on its surface. The effects of impingement of a gas jet on a liquid surface have been studied in detail in different scientific and industrial fields [19–21]. As described in previous work [1], the impact of the turbulent front upon the liquid surface causes a variation of the dimension of the dimple in time.

4. Discussion

According to the results, the influence of the electrical properties of the material on the discharge characteristics of an impinging plasma jet is clearly visible. Materials with very low conductivity, such as the dielectric PVC plate ($\sim10^{-17}$ µS/cm), favor the accumulation of charges on their surface. The total charge accumulated increases with the capacitance of the target [22] and, in our case, was significantly higher for the dielectric case in comparison with the metal and liquid substrates. The charge accumulation in turn induces the development of a radial electric field driving the formation of a SIW [11]. This results in the symmetrical radial expansion over the surface of the discharge, as clearly visible in Figure 4. SIW phenomenon is undetectable in the metal and the liquid cases, due to their much higher conductivity ($\sim3.7 \times 10^{11}$ µS/cm for the metal and 119 µS/cm for the liquid) that prevents the accumulation of charges on the surface.

On the other hand, the conductivity of the substrate also limits the power deposited by the plasma discharge. After the impinging of the ionization front on the substrate surface, a highly conductive channel connects the high-voltage electrode inside the jet source to the grounded substrate [10]. Once this connection occurs, the current passing through the plasma channel is limited by the conductivity of the target and the capacitance associated with it. As demonstrated by the power measurements (Table 1), the metal substrate, with a conductivity several order of magnitude higher than the other samples, recorded the highest dissipated power, albeit the values of measured power were in the same order of magnitude for all investigated cases. Moreover, although the liquid, with a higher conductivity, presented the lowest power, we hypothesize that other phenomena and factors, such as the target capacitance (higher for the dielectric), could have played a key role in the frame of ns-discharge interacting with a target. As was visible in the dielectric case, the plasma discharge propagated over the target surface and the emission intensity increased to t = 20 ns when the voltage pulse reached its peak. This suggests that the dielectric substrate, due to its capacitance, was still charging and, therefore, only partially limiting the current flow in the conductive channel. The liquid substrate on the other hand, characterized by a conductivity much lower than the metal (nearly 11 orders of magnitude) and a capacitance much lower than the dielectric, presented a singular behavior; no SIW was formed and there was a limitation of the deposited power. It must be noted that the liquid substrate is also the only substrate among those investigated to induce a partial modification of the surrounding gas composition due to evaporation. As reported by Ji et al. [23] in a similar configuration, the presence of a liquid film on a target presents a reduced impact on the characteristics of the plasma discharge propagation and intensity, while on the other hand, it can greatly enhance the production of OH radicals in close proximity to the target surface.

Concerning the plasma-induced alteration of the gas flow, the turbulent front could be clearly seen in all considered cases, but with different characteristics. In the case of the metal substrate, a turbulent alteration of the gas flow was visible from the discharge event (frame 0 ms in Figure 5). Later, a turbulent front propagated downstream with a velocity comparable with the gas flow's, as already observed in similar conditions in previous work [1]. Since the total expansion of the turbulent front was sufficiently fast enough to observe a re-established laminar He flow before the ignition of the next discharge event, we can confirm that the turbulent front phenomenon is directly associated with a combination of plasma-induced pressure waves and local electrode heating [1,24,25]. Moreover, because the Schlieren images (Figure 5) clearly showed that the magnitude of the turbulent structures varied between the investigated cases, we can hypothesize that turbulent front formation and its propagation is affected by the target characteristics. It may be speculated that there is a relation between turbulences and the target modification of the electric field distribution between the high-voltage electrode inside the plasma source and the targets [23]. In virtue of their conductivities and connections to ground, the metal and liquid targets greatly affected the electric field distribution, while the dielectric PVC target had a relative permittivity comparable to that of air ($\varepsilon_{r-PVC} = 3$), thus having a limited impact.

5. Conclusion

The present study highlights, through an ICCD and Schlieren imaging analysis, how the plasma plume changes its morphology and its light intensity as a consequence of the physical properties of the target, and how the fluid-dynamic of the plasma-induced turbulent front is affected by the substrate's nature. The proposed comparison between metal, dielectric and water substrates highlights significant differences between the associated plasma discharges. The electrical characteristics of the substrates may influence not only the plasma discharge propagation and ionization degree, but also the gas flow dynamics. Nanosecond pulsed plasma, interacting with a liquid substrate, presents a singular behavior that is not easily located midway between highly conductive and dielectric materials. Thus, future analysis should be directed to the investigation of substrate materials covering a wider range of conductivities, especially in the range of semiconductive materials. This would certainly help to better identify possible trends and provide better explanations. Concerning the transient turbulence generated in the gas flow by the plasma discharge, it has highlighted a dependence on the target nature. This aspect should be taken into account during plasma treatments, since it may affect through mixing of the gas composition in the plasma region, and may induce pressure fluctuations on the target surface. These aspects may hinder the uniformity of treatments, especially on complex and non-homogeneous surfaces like biological tissues.

The results of this work aim to be considered as one step closer to a full understanding of the complex interaction of non-equilibrium plasma with a substrate.

Author Contributions: Conceptualization, E.S., A.S., M.G. and V.C.; Data curation, E.S. and A.S.; Funding acquisition, V.C.; Investigation, E.S., A.S. and M.B.; Methodology, E.S., A.S. and M.B.; Project administration, V.C.; Resources, V.C.; Supervision, M.G. and V.C.; Writing—original draft, E.S., A.S. and V.C.; Writing—review & editing, E.S., A.S., M.G. and V.C.

Funding: This research received no external funding.

Acknowledgments: The authors would like to acknowledge Romolo Laurita for his contribution in liquid solution preparation and Eng. Alina Bisag for 3D rendering of the experimental setups.

Conflicts of Interest: The authors declare no conflicts of interest.

References

1. Stancampiano, A.; Simoncelli, E.; Boselli, M.; Colombo, V.; Gherardi, M. Experimental investigation on the interaction of a nanopulsed plasma jet with a liquid target. *Plasma Sources Sci. Technol.* **2018**, *27*, 12. [CrossRef]
2. Boselli, M.; Colombo, V.; Ghedini, E.; Gherardi, M.; Laurita, R.; Liguori, A.; Sanibondi, P.; Stancampiano, A. Schlieren high-speed imaging of a nanosecond pulsed atmospheric pressure non-equilibrium plasma jet. *Plasma Chem. Plasma Process.* **2014**, *34*, 853–869. [CrossRef]

3. Robert, E.; Sarron, V.; Riès, D.; Dozias, S.; Vandamme, M.; Pouvesle, J.-M. Characterization of pulsed atmospheric-pressure plasma streams (PAPS) generated by a plasma gun. *Plasma Sources Sci. Technol.* **2012**, *21*, 034017. [CrossRef]

4. Darny, T.; Pouvesle, J.-M.; Fontane, J.; Joly, L.; Dozias, S.; Robert, E. Plasma action on helium flow in cold atmospheric pressure plasma jet experiments. *Plasma Sources Sci. Technol.* **2017**, *26*, 11. [CrossRef]

5. Guaitella, O.; Sobota, A. The impingement of a kHz helium atmospheric pressure plasma jet on a dielectric surface. *J. Phys. D. Appl. Phys.* **2015**, *48*, 255202. [CrossRef]

6. Gazeli, K.; Bauville, G.; Fleury, M.; Jeanney, P.; Neveu, O.; Pasquiers, S.; Santos Sousa, J. Effect of the gas flow rate on the spatiotemporal distribution of Ar(1s5) absolute densities in a ns pulsed plasma jet impinging on a glass surface. *Plasma Sources Sci. Technol.* **2018**, *27*, 6. [CrossRef]

7. Yuanfu, Y.; Xuekai, P.; Dogan, G.; Fan, W.; Shuqun, W.; Xinpei, L.; Yue, Y.; Pei, X.; Gidon, D.; Wu, F. Investigation of plasma dynamics and spatially varying O and OH concentrations in atmospheric pressure plasma jets impinging on glass, water and metal substrates. *Plasma Sources Sci. Technol.* **2018**, *27*, 11.

8. Stancampiano, A.; Chung, T.-H.; Dozias, S.; Pouvesle, J.-M.; Mir, L.M.; Robert, E. Mimicking of human body electrical characteristic for easier translation of plasma biomedical studies to clinical applications. *IEEE Trans. Radiat. Plasma Med. Sci.* **2019**, in press. [CrossRef]

9. Judée, F.; Vaquero, J.; Guégan, S.; Fouassier, L.; Doufour, T. Atmospheric pressure plasma jets applied to cancerology: Correlating electrical configuration with in vivo toxicity and therapeutic efficiency. *J. Phys. D Appl. Phys* **2019**, *52*, 24. [CrossRef]

10. Darny, T.; Pouvesle, J.M.; Puech, V.; Douat, C.; Dozias, S.; Robert, E. Analysis of conductive target influence in plasma jet experiments through helium metastable and electric field measurements. *Plasma Sources Sci. Technol.* **2017**, *26*, 045008. [CrossRef]

11. Norberg, S.A.; Johnsen, E.; Kushner, M.J. Helium atmospheric pressure plasma jets touching dielectric and metal surfaces. *J. Appl. Phys.* **2015**, *118*, 013301. [CrossRef]

12. Gherardi, M.; Puač, N.; Marić, D.; Stancampiano, A.; Malović, G.; Colombo, V.; Petrović, Z.L. Practical and theoretical considerations on the use of ICCD imaging for the characterization of non-equilibrium plasmas. *Plasma Sources Sci. Technol.* **2015**, *24*, 064004. [CrossRef]

13. Robert, E.; Sarron, V.; Darny, T.; Riès, D.; Dozias, S.; Fontane, J.; Joly, L.; Pouvesle, J.-M. Rare gas flow structuration in plasma jet experiments. *Plasma Sources Sci. Technol.* **2014**, *23*, 012003. [CrossRef]

14. Riès, D.; Dilecce, G.; Robert, E.; Ambrico, P.F.; Dozias, S.; Pouvesle, J.-M. LIF and fast imaging plasma jet characterization relevant for NTP biomedical applications. *J. Phys. D. Appl. Phys.* **2014**, *47*, 275401. [CrossRef]

15. Lukes, P.; Dolezalova, E.; Sisrova, I.; Clupek, M. Aqueous-phase chemistry and bactericidal effects from an air discharge plasma in contact with water: Evidence for the formation of peroxynitrite through a pseudo-second-order post-discharge reaction of H_2O_2 and HNO_2. *Plasma Sources Sci. Technol.* **2014**, *23*, 015019. [CrossRef]

16. Laurita, R.; Barbieri, D.; Gherardi, M.; Colombo, V.; Lukes, P. Chemical analysis of reactive species and antimicrobial activity of water treated by nanosecond pulsed DBD air plasma. *Clin. Plasma Med.* **2015**, *3*, 53–61. [CrossRef]

17. Stancampiano, A.; Chung, T.; Dozias, S.; Pouvesle, J.; Mir, L.M.; Robert, E. To ground or not to ground? That is a key question during plasma medical treatment. In Proceedings of the ISPC 24, Naples, Italy, 9–14 June 2019.

18. Traldi, E.; Boselli, M.; Simoncelli, E.; Stancampiano, A.; Gherardi, M.; Colombo, V.; Settles, G.S. Schlieren imaging: A powerful tool for atmospheric plasma diagnostic. *EPJ Tech. Instrum.* **2018**, in press. [CrossRef]

19. Bruggeman, P.J.; Kushner, M.J.; Locke, B.R.; Gardeniers, J.G.E.; Graham, W.G.; Graves, D.B.; Hofman-Caris, R.C.H.M.; Maric, D.; Reid, J.P.; Ceriani, E.; et al. Plasma—Liquid interactions: A review and roadmap. *Plasma Sources Sci. Technol.* **2016**, *25*, 5. [CrossRef]

20. Norberg, S.A.; Tian, W.; Johnsen, E. Helium atmospheric pressure plasma jets interacting with wet cells: Delivery of electric fields. *J. Phys. D Appl. Phys* **2016**, *49*, 16. [CrossRef]

21. Norberg, S.A.; Tian, W.; Johnsen, E.; Kushner, M.J. Atmospheric pressure plasma jets interacting with liquid covered tissue: Touching and not-touching the liquid. *J. Phys. D. Appl. Phys.* **2014**, *47*, 475203. [CrossRef]

22. Ito, Y.; Fukui, Y.; Urabe, K.; Sakai, O.; Tachibana, K. Effect of Series Capacitance and Accumulated Charge on a Substrate in a Deposition Process with an Atmospheric-Pressure Plasma Jet. *Jpn. J. Appl. Phys.* **2010**, *49*, 066201. [CrossRef]
23. Ji, L.; Yan, W.; Xia, Y.; Liu, D. The effect of target materials on the propagation of atmospheric-pressure plasma jets. *J. Appl. Phys.* **2018**, *123*. [CrossRef]
24. Zhang, S.; Sobota, A.; van Veldhuizen, E.M.; Bruggeman, P.J. Gas flow characteristics of a time modulated APPJ: The effect of gas heating on flow dynamics. *J. Phys. D. Appl. Phys.* **2015**, *48*, 015203. [CrossRef]
25. Mitsugi, F.; Kusumegi, S.; Kawasaki, T.; Nakamiya, T.; Sonoda, Y. Detection of Pressure Waves Emitted From Plasma Jets With Fibered Optical Wave Microphone in Gas and Liquid Phases. *IEEE Trans. Plasma Sci.* **2016**, *44*, 3077–3082. [CrossRef]

 *plasma*

Article

Hydrogen Peroxide Interference in Chemical Oxygen Demand Assessments of Plasma Treated Waters

Joseph Groele [1],* and John Foster [2]

[1] Department of Mechanical Engineering, University of Michigan, Ann Arbor, MI 48109, USA
[2] Department of Nuclear Engineering and Radiological Sciences, University of Michigan, Ann Arbor, MI 48109, USA
* Correspondence: jrgroele@umich.edu; Tel.: +1-716-307-5651

Received: 16 May 2019; Accepted: 2 July 2019; Published: 5 July 2019

Abstract: Plasma-driven advanced oxidation represents a potential technology to safely re-use waters polluted with recalcitrant contaminants by mineralizing organics via reactions with hydroxyl radicals, thus relieving freshwater stress. The process results in some residual hydrogen peroxide, which can interfere with the standard method for assessing contaminant removal. In this work, methylene blue is used as a model contaminant to present a case in which this interference can impact the measured chemical oxygen demand of samples. Next, the magnitude of this interference is investigated by dosing de-ionized water with hydrogen peroxide via dielectric barrier discharge plasma jet and by solution. The chemical oxygen demand increases with increasing concentration of residual hydrogen peroxide. The interference factor should be considered when assessing the effectiveness of plasma to treat various wastewaters.

Keywords: non-equilibrium plasma applications; water treatment; advanced oxidation

---

1. Introduction

Demand for freshwater is rapidly rising due to expanding agriculture and industrialization; meanwhile, population growth, changing climate, accidental spills, and deteriorating infrastructure further exacerbates the problem of freshwater scarcity [1]. One possible approach to managing this impeding crisis is to re-use wastewater by removing contaminants; however, wastewaters contain contaminants of emerging concern (CECs) which are not readily removed by traditional water treatment processes. These CECs include pharmaceuticals, industrial chemicals such as per- and polyfluoroalkyl substances (PFAS), pesticides for agriculture, and microcystins [2–6]. Fortunately, these CECs can be removed from wastewaters using advanced oxidation processes [7].

Advanced oxidation processes (AOPs) are a category of chemical treatment methods for removing persistent organic pollutants from waters and wastewaters via reactions with highly reactive oxidizing agents, namely hydroxyl radicals. Traditional AOPs typically leverage combinations of ozone, hydrogen peroxide, and ultraviolet light to facilitate in-situ generation of hydroxyl radicals for non-selective decomposition of organic contaminants [8,9]. Through a series of chain reactions known as mineralization, hydroxyl radicals react with organics to ultimately form carbon dioxide, water, and mineral ions, with aldehydes and carboxylic acids often serving as intermediate breakdown products [10].

Plasma-liquid interaction can generate hydrogen peroxide, ozone, and other reactive species along with ultraviolet radiation and short-lived highly reactive hydroxyl radicals, thus representing a novel AOP for the removal of recalcitrant organic compounds from waters and wastewaters [11–13]. The formation of these reactive species is initiated primarily through energetic electrons from the plasma region colliding with atoms or molecules, either in the gas phase or at the liquid surface.

Subsequent production of reactive species can also be facilitated by recombination of radicals and de-excitation of metastables [14].

Plasma discharges in contact with water can generate multitudes of reactive oxygen species, including superoxide, hydroperoxyl, and atomic oxygen. If the discharge is in air, then reactive nitrogen species will also form, including nitric oxide, nitrite, nitrate, and peroxynitrate [15]. While all of these reactive oxygen and nitrogen species may contribute to degradation of contaminants in wastewaters, the most important species for advanced oxidation are hydroxyl radicals, hydrogen peroxide, and ozone. In particular, hydroxyl radicals react non-selectively with organics, including contaminant intermediate products such as short chain carboxylic acids, to allow for complete mineralization of most organic pollutants [16].

The production of hydroxyl radicals in plasmas in or in contact with water can occur through a number of different pathways, with the relative importance of these pathways depending on the plasma parameters and the gas composition [17]. Some of these hydroxyl radicals produced in the discharge diffuse to the liquid surface and are transported across the gas-water interface to either react with pollutants in the water or scavenge themselves to form longer lived hydrogen peroxide [12,18].

In the bulk liquid, hydrogen peroxide can react with any dissolved ozone from the discharge to produce more hydroxyl radicals (i.e., peroxone process [19]) for in-situ oxidation of organics, even after the plasma has been turned off. However, not all of the hydrogen peroxide is consumed through reactions with ozone, and a portion of the post-discharge hydroxyl radicals produced by the peroxone process will dimerize back into hydrogen peroxide. In this way, plasma-driven advanced oxidation can leave residual hydrogen peroxide in treated waters that may persist for days, as can also happen with traditional AOPs [20].

The goal of plasma-driven advanced oxidation, and AOPs in general, is to chemically remove contaminants from waters. The five most prominent methods for assessing contaminant removal are liquid chromatography tandem mass spectrometry (LC-MS), spectrophotometry, total organic carbon (TOC), biochemical oxygen demand (BOD), and chemical oxygen demand (COD) [21–24]. Spectrophotometry and LC-MS allow for the determination of specific species concentrations, whereas TOC, BOD, and COD are non-specific water quality parameters that provide a measurement of the overall pollution potential of a water sample. When initially investigating the performance of AOPs for removal of contaminants from wastewaters, LC-MS should be used to validate results from other methods; however, for general monitoring of effluent quality in water and wastewater treatment processes, TOC, BOD, and COD are more commonly used due to the simplicity of the test procedure and easily interpreted results.

As the name suggests, TOC is a measure of the total organic carbon contained in a water sample. This measurement involves two stages: total carbon (TC) and inorganic carbon (IC) measurements, with the difference providing the TOC of the sample. First, combustion catalytic oxidation of the sample converts the organic carbon to carbon dioxide, which is then cooled and humidified before being detected by an infrared gas analyzer to measure the TC content of the sample. The second step involves acidifying the sample to pH < 4, making bicarbonate and carbonate unstable [21,25]. Thus, the IC in the sample is converted to carbon dioxide, either free $CO_{2(aq)}$ or in the form of carbonic acid, that can be purged with a CO_2-free gas to isolate the carbon dioxide, which is subsequently cooled, humidified, and detected by the gas analyzer as the IC measurement.

While TOC measurements focus on the carbon content of a sample directly, BOD and COD are measurements of the amount of oxygen consumed in the complete oxidation of organics to carbon dioxide, water, and mineral salts. As such, BOD and COD are particularly well suited for monitoring water treatment processes involving oxidation of organics and for informing the design of the water treatment process (e.g., how much oxidant must be used to remove the contaminants present in the waste stream).

More specifically, BOD is a measure of the oxygen consumed by bacteria in the oxidation of organics and is representative of contaminant decomposition in the natural environment, making this method most suitable for predicting the effects of the organic contaminants on the dissolved oxygen levels of receiving waters. In contrast, COD is a measure of the oxygen consumed in the chemical oxidation of organics [23]. The BOD test typically takes five days to get results, meanwhile the COD test can provide results in less than three hours.

Due to the short analysis time and ability to correlate with BOD, the COD test has become the industry standard for rapid and frequent monitoring of water treatment process efficiency and effluent quality. The most common method is the closed reflux, colorimetric method with potassium dichromate oxidizing agent [22]. Pre-formulated commercial reagent mixtures are available for rapid COD analysis and provide consistent results between different laboratories. The problem with COD assessments of plasma treated wastewaters is that residual peroxide interferes with the measurement. Indeed, this interference is an issue for any AOP involving residual peroxide [20]. This paper first discusses a case in which this interference can lead to misinterpretation of plasma treatment results, and then investigates the magnitude of the interference in COD assessments of waters containing residual hydrogen peroxide.

2. Materials and Methods

2.1. Underwater Dielectric Barrier Discharge Plasma Jet

The discharge configuration used to bring non-equilibrium plasma in contact with water in this investigation is an underwater dielectric barrier discharge (DBD) plasma jet. The set-up is shown in Figure 1 and is based on the design by Foster et al. [26]. The plasma applicator consists of an 18 gauge tin-plated copper high-voltage wire centered coaxially within a quartz tube with 4 mm ID, 6.35 mm OD, and 150 mm length, and a tin-plated copper wire coil wrapped around the outside of the quartz tube to serve as the grounded electrode. In this case, the quartz acts as the dielectric barrier preventing a thermal arc from forming.

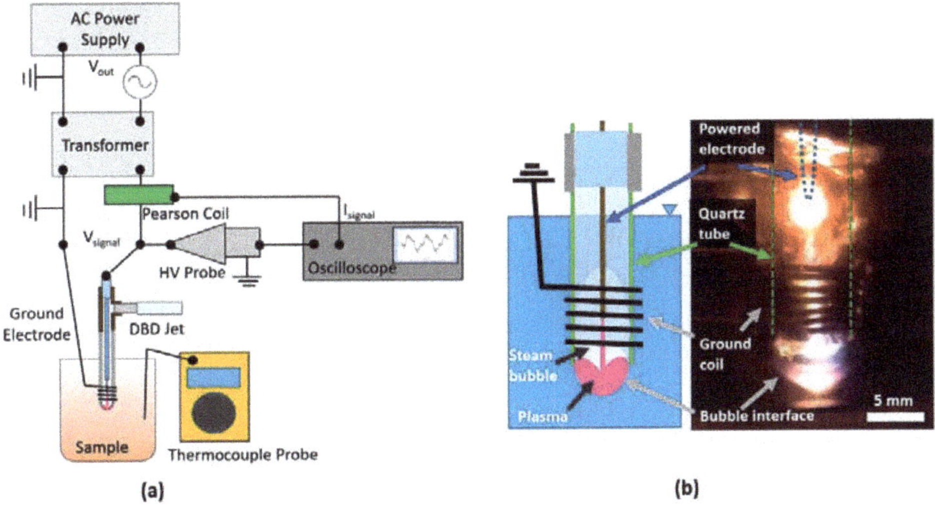

Figure 1. (**a**) Schematic of the DBD plasma jet experimental set-up used in this work. (**b**) Schematic and image of the DBD plasma jet apparatus operating in steam-mode discharge in de-ionized water.

The central high-voltage electrode is excited with a 5 kHz sinusoidal voltage at 4 kV$_{p-p}$, provided by an Elgar model 501SL power supply (AMETEK, Inc., Berwyn, PA, USA) with a 50:1 step-up transformer (SP-225 Plasma Technics, Inc., Racine, WI, USA). Voltage was measured using a high-voltage probe (P6015A, Tektronix, Beaverton, OR, USA), and the discharge current was measured with a Pearson coil current monitor (6585, Pearson Electronics, Inc., Palo Alto, CA, USA). Current and voltage waveforms were recorded using a 2 GHz oscilloscope (Wavepro 7200a, Teledyne LeCroy, Chestnut Ridge, NY, USA). Typical current-voltage data is shown in Figure 2. Data from previous Lissajous experiments by Garcia et al. with the DBD plasma jet operating in steam-mode at these power supply conditions indicate that about 56 W of power are deposited in the discharge, with peak steam temperatures around 2800 K, as determined from comparing theoretical simulation of OH(A-X) spectra with experimental optical emission spectra [27].

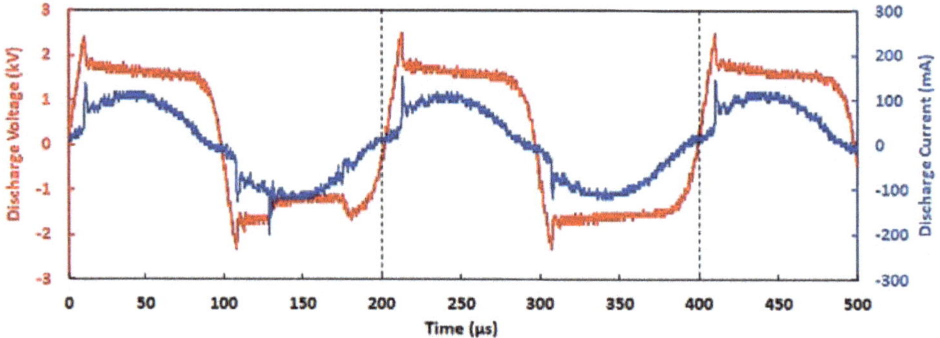

Figure 2. Discharge voltage and current signals measured during operation of the underwater DBD plasma jet operating in steam-mode at 5 kHz.

This DBD plasma jet can be operated with or without gas flow. In this work, the plasma jet operates without any input gas. This so-called "steam mode" discharge relies on local evaporation near the tip of the high-voltage electrode to form a vapor bubble that acts as the low-density region for plasma formation. Although a portion of the deposited power goes into heating and evaporating water, operating in steam-mode requires no consumables and minimizes the production of NO_x, as evidenced from optical emission spectroscopy [28,29], which can interfere with the iodometric titration method for quantifying hydrogen peroxide concentration, discussed in Section 2.3.

2.2. Sample Preparation

Samples of de-ionized water (Milli-Q EMD Millipore, Burlington, MA, USA, electrical conductivity ≈ 3 µS/cm) are dosed with hydrogen peroxide either by solution (30%, Fisher Chemical, Hampton, NH, USA) using a variable volume sampler pipette (Thermo Fisher Scientific, Waltham, MA, USA, resolution: 1 µL) or by underwater DBD plasma jet. When dosing with the DBD plasma jet, the opening of the quartz tube is placed approximately in the center of a 50 mL sample of de-ionized water and power is provided to the central high-voltage electrode. Approximately one second is required for the vapor bubble to form at the electrode tip in which the plasma discharge subsequently develops.

Plasma is applied to the sample via the DBD plasma jet for 30 s at a time, and the increase in sample temperature throughout the treatment duration is measured using a Fluke 87V true RMS multimeter with a type-K thermocouple (Fluke Corporation, Everett, WA, USA, resolution: 0.1 °C, accuracy: 0.05%). After 30 s of treatment, the sample is placed in an ice bath to cool down to room temperature before further treatment. The objective was to keep the bulk liquid sample below 60 °C. After treatment, the samples are stored in a dark cabinet for approximately 20 h before the hydrogen peroxide concentration and COD are measured to allow post-discharge reactions to take place while mitigating hydrogen peroxide photo-dissociation.

Samples of methylene blue (MB) are prepared by dissolving solid MB powder (M291-25 Fisher Chemical) in de-ionized water (Milli-Q EMD Millipore). A low concentration sample of 5 mg/L and a high concentration sample of 1000 mg/L are prepared and treated with the underwater DBD plasma jet to demonstrate the hydrogen peroxide interference problem, discussed in Section 3. Plasma treatment of the MB samples follows the same procedure as described for dosing de-ionized water samples with hydrogen peroxide by DBD plasma jet, including monitoring the temperature to keep the sample below 60 °C. Again, treated MB samples were stored in a dark cabinet for 20 h before assessing the COD.

2.3. Hydrogen Peroxide Measurement

The hydrogen peroxide concentration in the sample is quantified using the iodometric titration method (Hach test kit model HYP-1, Hach Company, Loveland, CO, USA, resolution: 1 mg/L H_2O_2) immediately prior to COD assessment. The titration method involves the oxidation of iodide to iodine (Equation (1)) in the presence of a molybdate catalyst and a starch indicator. The starch indicator forms a dark blue complex with iodine. Subsequent titration with thiosulfate under acidic conditions (approximate pH of 4) converts the iodine back to iodide (Equation (2)), and the color change from blue to transparent indicates the titration endpoint.

$$H_2O_2 + 2KI + H_2SO_4 \rightarrow I_2 + K_2SO_4 + 2H_2O \tag{1}$$

$$I_2 + 2Na_2S_2O_3 \rightarrow Na_2S_4O_6 + 2NaI \tag{2}$$

$$2I^- + 2NO_2^- + 4H^+ \rightarrow I_2 + 2NO + 2H_2O \tag{3}$$

This iodometric titration method is subject to interference from nitrite ions (Equation (3)); thus, nitrogen species from air plasma discharges will interfere with this measurement. To mitigate the generation of reactive nitrogen species that could result in aqueous nitrite, the DBD plasma jet is operated in steam mode, as discussed in Section 2.1. For future studies, colorimetric assay by titanyl sulfate reagent with azide for elimination of nitrite interference is a preferred method of hydrogen peroxide quantification due to superior selectivity, making it suitable for hydrogen peroxide concentration measurements in waters treated with air plasmas [15,30].

2.4. Chemical Oxygen Demand Measurement

The assessment of COD in this work was performed using the USEPA 4.10.4 approved method, which is the closed reflux, colorimetric method with potassium dichromate oxidizer. Pre-formulated reagent was purchased from Hanna Instruments (HI9375A-25 COD reagent low range: 0 to 150 mg/L as O_2, accuracy: ±5 mg/L or ±4% of reading at 25 °C, resolution: 1 mg/L) along with a Hanna Instruments photometer (HI83399, Hanna Instruments, Woonsocket, RI, USA) for measuring sample absorbance using an LED light source with narrow band interference filter at 420 nm.

3. Results and Discussion

To illustrate a picture of the problem, a 100 mL sample of 5 mg/L MB is prepared and the COD is assessed before and after treatment. The results are shown in Table 1. In general, the COD is expected to decrease with plasma treatment time as the MB molecules are mineralized. Indeed, the color of the dye disappears after 15 min of treatment indicating the destruction of MB, as seen in Figure 3a; yet, the measured COD actually increases with treatment time. No additional organics are being added to the solution by the plasma; thus, one of the plasma-derived species being transported to the liquid during treatment must be contributing to the measured COD of the sample.

Table 1. Measured COD values for untreated and treated samples of MB. MB-A samples (left) contain initial MB concentration of 5 mg/L while MB-B samples (right) start with 1000 mg/L MB.

MB-A Sample	COD (mg/L)	MB-B Sample	COD (mg/L)
Untreated (5 mg/L)	6 ± 5	Untreated (1000 mg/L)	1680 ± 75
Treated for 7 min	16 ± 5	Treated for 40 min	99 ± 5
Treated for 15 min	30 ± 5	Treated for 120 min	62 ± 5

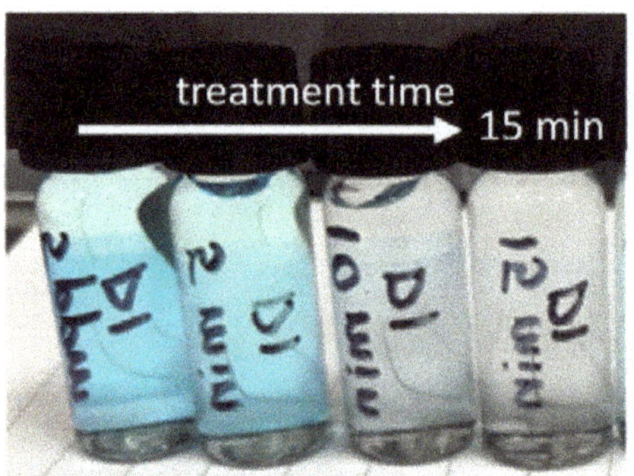

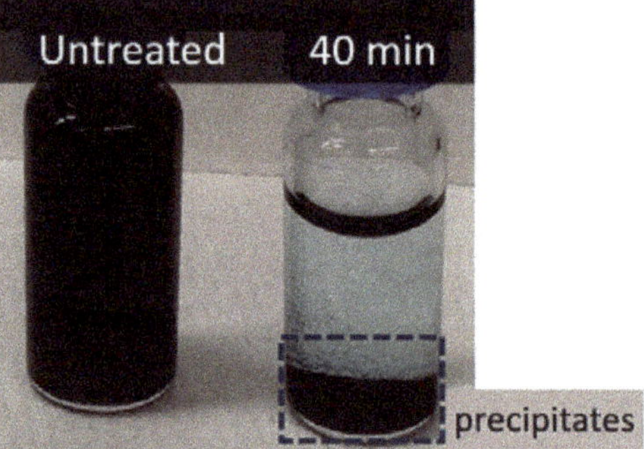

Figure 3. (**a**) Images of samples starting with 5 mg/L MB showing decrease in color indicating destruction of the MB dye. (**b**) Image of untreated sample with 1000 mg/L MB and sample after 40 min of plasma treatment showing MB precipitates and mostly clear supernatant.

The two primary long-lived plasma-derived species that could be present in the sample after plasma treatment are hydrogen peroxide and ozone, since nitrates and nitrites should not be generated under steam-mode operation. Of these, hydrogen peroxide has been found to interfere with the standard COD assessment by reducing the potassium dichromate oxidizing agent according to the overall oxidation reaction given in Equation (4) [31]. From this equation, the theoretical hydrogen peroxide interference is calculated to be 470.6 mg of COD as O_2 per 1000 mg H_2O_2.

$$K_2Cr_2O_7 + 3H_2O_2 + 4H_2SO_4 \rightarrow K_2SO_4 + Cr_2(SO_4)_3 + 7H_2O + 3O_2 \tag{4}$$

Researchers investigating plasma treatment of wastewaters must be particularly careful when interpreting COD results because the interference from residual hydrogen peroxide may not be evident. For example, two 100 mL samples of 1000 mg/L MB were prepared and treated with the underwater DBD plasma jet. After treatment, the COD decreased from the initial value of 1680 ± 75 mg/L, as shown in Table 1, and the supernatant becomes transparent as oxidized MB precipitates out of the solution, as shown in Figure 3b. This case of high initial COD relative to the residual peroxide concentration masks the peroxide interference effect. Still, any residual peroxide left in the sample after treatment will contribute to the COD. Hence, the COD contribution from organics should be less than the 99 mg/L after 40 min of treatment indicated in Table 1, since there is some contribution from residual peroxide.

In this case of high initial dye concentration (MB-B Sample, Table 1), the residual peroxide contribution to the COD may have been up 99 mg/L, but since the COD decreases from 1680 mg/L to 99 mg/L, it appears that the process is working as intended (COD decreases as MB is oxidized). In this case, the COD of the residual peroxide is at least 17 times less than the COD of the organics present in the sample before treatment, thus the interference effect is not immediately apparent. In reality, the treatment process is likely working better (faster organic removal rate) than would be indicated by the COD test because of the interference from residual peroxide. However, in the case of low initial dye concentration (MB-A Sample, Table 1), the measured COD increases with plasma treatment time because the COD contribution from residual peroxide after treatment is greater than the COD of the organics present in the sample before treatment. By simply looking at the measured COD before and after treatment, it would appear as if the process did not work (i.e., organics were not removed), because the peroxide interference has not been corrected for.

To investigate the magnitude of hydrogen peroxide interference in COD assessments, de-ionized water was dosed with hydrogen peroxide both by solution and by DBD plasma jet according to the procedure described in Section 2.2. The hydrogen peroxide concentration is measured immediately before COD assessment. In both cases, the COD increases with increasing hydrogen peroxide concentration, as shown in Figure 4, even though there were no organics present in the samples. This proportional relationship was also found by Lee et al. for conventional advanced oxidation processes, although the interference magnitude varied depending on the water quality and treatment process [20].

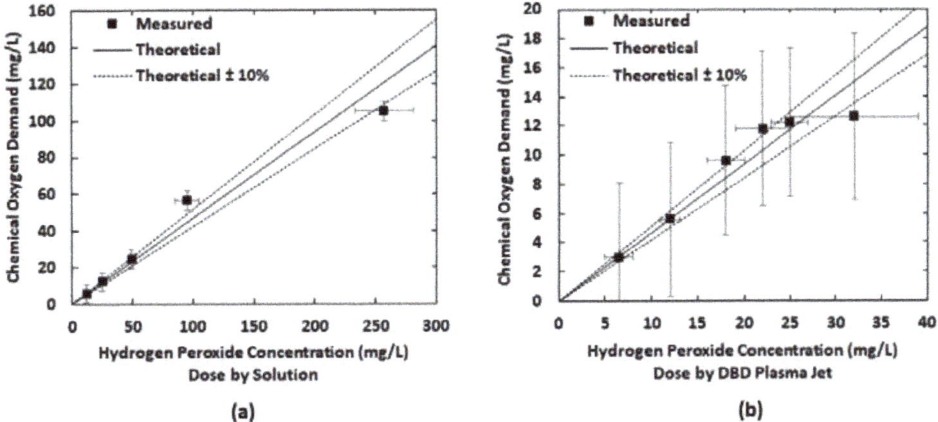

Figure 4. (a) COD vs. H_2O_2 concentration, where H_2O_2 is added to samples by a solution of 30% H_2O_2. (b) COD vs. H_2O_2 concentration, where H_2O_2 is added to samples by exposure to the underwater DBD plasma jet.

These results seem to confirm that the measured COD results from residual peroxide reacting with the potassium dichromate oxidizing agent during the COD assessment. The magnitude of the interference found in this experiment is approximately equal to the theoretical value from Equation (4), but this value can vary in practice by up to 16% percent depending on the specific organic compounds and oxidizing species that are present in the sample being assessed, particularly in real wastewaters [20]. When AOPs are used to remove trace levels of contaminants from relatively low COD waters, as can occur at final disinfection stages of drinking water treatment plants, the uncertainty in the actual interference magnitude can lead to misinterpretation of results. While reagents may be used to quench residual hydrogen peroxide in attempt to remove this interference, these reagents often introduce their own interferences with the COD assessment [31]. More research is needed to investigate how the residual oxidants and other long-lived species from plasma treatment affect the COD measurement in various synthetic and real wastewaters in order to develop correction methods for appropriate interpretation of results.

4. Conclusions

Wastewater treatment plants use COD to measure treatment process efficiency and quality of treated water. Plasma-water interaction produces residual peroxide which interferes with the COD measurement. As a result, COD assessments of plasma treated waters will show incomplete removal of oxygen demand and may be viewed as a disadvantage of plasma-based advanced oxidation for municipal wastewater treatment plants and companies looking to treat their wastewater streams. Indeed, when researching new wastewater treatment technologies, LC-MS should be used to determine residual organic content before applying standard water quality diagnostics. This paper therefore describes precautions necessary for plasma practitioners to take into account so as to yield credible contaminant decomposition measurements. Plasma in contact with water produces a host of reactive oxygen species, and so caution is necessary when interpreting the results.

Author Contributions: Investigation, analysis, and writing, J.G.; conceptualization, resources, and supervision, J.F.

Funding: This research was funded by the National Science Foundation (NSF 1700848) and the U.S. Department of Energy (DOE DE-SC0001939).

Conflicts of Interest: The authors declare no conflicts of interest.

References

1. Mekonnen, M.M.; Hoekstra, A.Y. Four billion people facing severe water scarcity. *Sci. Adv.* **2016**, *2*, e1500323. [CrossRef] [PubMed]
2. Suarez, S.; Carballa, M.; Omil, F.; Lema, J.M. How are pharmaceutical and personal care products (PPCPs) removed from urban wastewaters. *Rev. Environ. Sci. Biotechnol.* **2008**, *7*, 125–128. [CrossRef]
3. Vandeivere, P.C.; Bianchi, R.; Verstraete, W. Treatment and reuse of wasatewater from the textile wet-processing industry: Review of emerging technologies. *J. Chem. Technol. Biotechnol.* **1998**, *72*, 289–302. [CrossRef]
4. Buck, R.C. Perfluoroalkyl and polyfluoroalkyl substances in the environment: Terminology, classification, and origins. *Integr. Environ. Assess. Manag.* **2011**, *7*, 513–514. [CrossRef] [PubMed]
5. Jiang, X.; Lee, S.; Mok, C.; Lee, J. Sustainable methods for decontamination of microcystin in water using cold plasma and uv with reusable TiO_2 nanoparticle coating. *Int. J. Environ. Res. Public Health* **2017**, *14*, 480–491. [CrossRef] [PubMed]
6. Zhang, H.; Huang, Q.; Ke, Z.; Yang, L.; Wang, X.; Yu, Z. Degradation of microcystin-LR in water by glow discharge plasma oxidation at the gas-solution interface and its safety evaluation. *Water Res.* **2012**, *46*, 6554–6562. [CrossRef] [PubMed]
7. Crook, J.; Surampalli, R.U. Water reclamation and reuse criteria in the U.S. *Water Sci. Tech.* **1996**, *33*, 451–462. [CrossRef]
8. Oturan, M.A.; Aaron, J. Advanced oxidation processes in water/wastewater treatment: Principles and applications. A review. *Crit. Rev. Environ. Sci. Technol.* **2014**, *44*, 2577–2641. [CrossRef]
9. Glaze, W.H.; Kang, J.; Chapin, D.H. The chemistry of water treatment processes involving ozone, hydrogen peroxide, and ultraviolet radiation. *Ozone Sci. Eng.* **1987**, *9*, 335–352. [CrossRef]
10. Santos, L.C.; Poli, A.L.; Cavalheiro, C.C.S.; Neumann, M.G. The UV/H_2O_2-photodegradation of poly(ethyleneglycol) and model compounds. *J. Braz. Chem. Soc.* **2009**, *20*, 1467–1472. [CrossRef]
11. Bruggeman, P.; Leys, C. Non-thermal plasmas in and in contact with liquids. *J. Phys. D Appl. Phys.* **2009**, *42*, 053001. [CrossRef]
12. Liu, Z.C.; Liu, D.X.; Chen, C.; Li, D.; Yang, A.J.; Rong, M.Z.; Chen, H.L.; Kong, M.G. Physicochemical processes in the indirect inderaction between surface air plasma and deionized water. *J. Phys. D Appl. Phys.* **2015**, *48*, 495201. [CrossRef]
13. Tian, W.; Tachibana, K.; Kushner, M.J. Plasmas sustained in bubbles in water: Optical emission and excitation mechanisms. *J. Phys. D Appl. Phys.* **2014**, *47*, 055202. [CrossRef]
14. Magureanu, M.; Bradu, C.; Parvulescu, V.I. Plasma processes for the treatment of water contaminated with harmful organic compounds. *J. Phys. D Appl. Phys.* **2018**, *51*, 313002. [CrossRef]
15. Lukes, P.; Dolezalova, E.; Sisrova, I.; Clupek, M. Aqueous-phase chemistry and bactericidal effects from an air discharge plasma in contact with water: Evidence for the formation of peroxynitrite through a pseudo-second-order post-discharge reaction of H_2O_2 and HNO_2. *Plasma Sources Sci. Technol.* **2014**, *23*, 015019. [CrossRef]
16. Lukes, P.; Locke, B.R. Plasmachemical oxidation processes in a hybrid gas-liquid electrical discharge reactor. *J. Phys. D Appl. Phys.* **2005**, *38*, 4074–4081. [CrossRef]

17. Bruggeman, P.; Schram, D.C. On OH production in water containing atmospheric pressure plasmas. *Plasma Sources Sci. Technol.* **2010**, *19*, 45025. [CrossRef]

18. Anderson, C.E.; Cha, N.R.; Linday, A.D.; Clark, D.S.; Graves, D.B. The role of interfacial reactions in determining plasma-liquid chemistry. *Plasma Chem. Plasma Processes.* **2016**, *36*, 1393–1415. [CrossRef]

19. Fischbacher, A.; Sonntag, J.; Sonntag, C.; Schmidt, T.C. The OH radical yield in the $H_2O_2 + O_3$ (peroxone) reaction. *Environ. Sci. Technol.* **2013**, *47*, 9959–9964. [CrossRef]

20. Lee, E.; Lee, H.; Kim, Y.K.; Sohn, K.; Lee, K. Hydrogen peroxide interference in chemical oxygen demand during ozone based oxidation of anaerobically digested livestock wastewater. *Int. J. Environ. Sci. Tech.* **2011**, *8*, 381–388. [CrossRef]

21. Volk, C.; Wood, L.; Johnson, B.; Robinson, J.; Zhu, H.W.; Kaplan, L. Monitoring dissolved organic carbon in surface and drinking waters. *J. Environ. Monit.* **2002**, *4*, 43–47. [CrossRef]

22. *Standard Methods for the Examination of Water and Wastewater*, 15th ed.; American Public Health Association, American Water Works Association, Water Environment Federation: Washington, DC, USA, 1999.

23. Sawyer, C.N.; McCarty, P.L.; Parkin, G.F. *Chemistry for Environmental Engineering and Science*, 5th ed.; McGraw-Hill: New York, NY, USA, 2003.

24. Norton, J.F. *Standard Methods for the Examination of Water and Sewage*, 9th ed.; American Public Health Association: Washington, DC, USA, 1946.

25. Sharp, J.H. Total organic carbon in seawater—Comparison of measurements using persulfate oxidation and high temperature combustion. *Mar. Chem.* **1973**, *1*, 211–229. [CrossRef]

26. Foster, J.E.; Weatherford, B.; Gillman, E.; Yee, B. Underwater operation of a DBD plasma jet. *Plasma Sources Sci. Technol.* **2010**, *19*, 025001. [CrossRef]

27. Garcia, M.C.; Gucker, S.N.; Foster, J.E. Understanding the plasma and power characteristics of a self-generated steam bubble discharge. *J. Phys. D Appl. Phys.* **2015**, *48*, 355203. [CrossRef]

28. Gucker, S.N.; Foster, J.E.; Garcia, M.C. An investigation of an underwater steam plasma discharge as alternative to air plasmas for water purification. *Plasma Sources Sci. Technol.* **2015**, *24*, 055005. [CrossRef]

29. Ni, G.; Zhao, G.; Jiang, Y.; Li, J.; Meng, Y.; Wang, X. Steam plasma jet treatment of phenol in aqueous solution at atmospheric pressure. *Plasma Process. Polym.* **2013**, *10*, 353–363. [CrossRef]

30. Eisenberg, G. Colorimetric determination of hydrogen peroxide. *Ind. Eng. Chem. Anal. Ed.* **1943**, *15*, 327–328. [CrossRef]

31. Talini, I.; Anderson, G.K. Interference of hydrogen peroxide on the standard COD test. *Wat. Res.* **1992**, *26*, 107–110. [CrossRef]

Article

Ignition of a Plasma Discharge Inside an Electrodeless Chamber: Methods and Characteristics

Mounir Laroussi

Electrical & Computer Engineering Department, Old Dominion University, Norfolk, VA 23529, USA; mlarouss@odu.edu; Tel.: +1-757-683-6369

Received: 31 July 2019; Accepted: 8 October 2019; Published: 14 October 2019

Abstract: In this paper the generation and diagnostics of a reduced pressure (300 mTorr to 3 Torr) plasma generated inside an electrodeless containment vessel/chamber are presented. The plasma is ignited by a guided ionization wave emitted by a low temperature pulsed plasma jet. The diagnostics techniques include Intensified Charge Coupled Device (ICCD) imaging, emission spectroscopy, and Langmuir probe. The reduced-pressure discharge parameters measured are the magnitude of the electric field, the plasma electron number density and temperature, and discharge expansion speed.

Keywords: plasma jet; electric field; plasma bullet; diffuse plasma; ionization wave

---

1. Introduction

Low temperature plasma jets exhibit large electric fields at the tip of their plasma plumes. The magnitude of this field was measured by several investigators and was found to be in the 10–30 kV/cm range [1–4]. Figure 1 shows measurements of the electric field reported by Begum et al. [1] while Figure 2 shows simulation results reported by Naidis [5]. When impinging on a dielectric material, the plasma plume causes charge build up on the surface of the dielectric surface. Therefore, via capacitive coupling, the electric field can be effectively transmitted to the area behind the dielectric barrier [6–8]. Under reduced pressure, the transmitted field can be large enough to start a discharge, which quickly propagates and grows in volume [9].

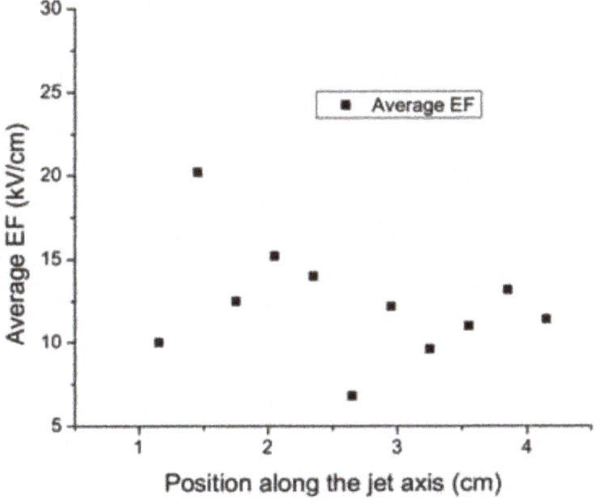

Figure 1. Mean magnitude of the electric field along the axis for a plasma plume emitted by a pulsed plasma jet [1].

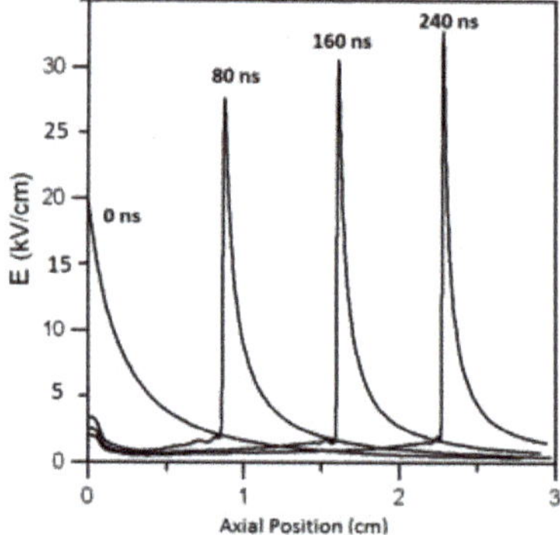

Figure 2. Electric field axial profile at different times for a helium plasma jet emerging into surrounding air. The jet radius is 0.25 cm. Figure adapted from Figure 1 of [5].

In this paper the characteristics of the plasma ignited inside an electrodeless dielectric chamber by an external plasma jet are reported. The chamber has no direct physical or electric connections to the externally applied plasma jet. The diagnostics techniques used, which include emission spectroscopy, fast imaging, and Langmuir probe, allow for the measurement of the magnitude of the transmitted electric field inside the chamber, the plasma electron number density and temperature, and discharge propagation speed.

2. Materials and Methods

Low temperature plasma jets launched in the ambient environment are enabled by guided ionization waves [10]. The electric field at the front of these waves can be quite large and can be transmitted across a dielectric barrier. If the dielectric barrier constitutes the wall of a chamber where the pressure can be controlled, then a diffuse plasma can be generated inside the chamber below a certain pressure threshold [9]. This principle was used here to generate a reduced pressure plasma inside a Pyrex cross-shaped tube that has no electrodes or electrical connections. Figure 3 shows the experimental setup and a photo of the reduced pressure plasma inside the Pyrex tubing.

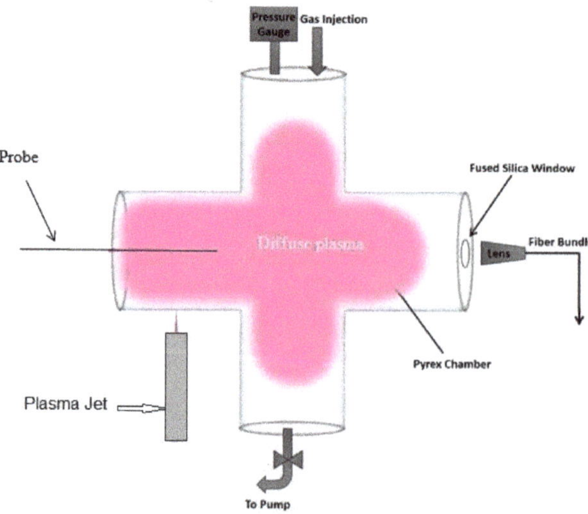

Figure 3. *Cont.*

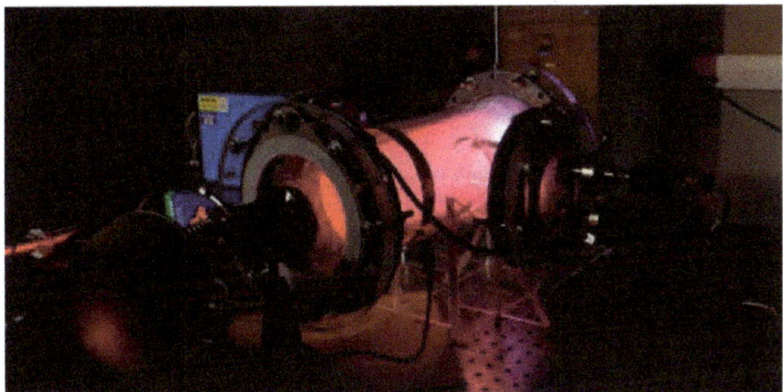

Figure 3. Schematic of the experimental setup (top) and photo of the large volume plasma inside the tubing/chamber (bottom). The plasma jet is applied externally on the far side of the chamber. The chamber is a Pyrex glass cross with an arm 18 inches long (inner diameter is 6 inches) and a second arm 16 inches long (inner diameter is 4 inches). For this photo, the pressure inside the chamber was 3 Torr and the gas was air.

The tubing/chamber is made of Pyrex glass, where the pressure can be reduced gradually down to the mTorr range. The plasma jet is physically independent of the chamber and is applied externally. It is driven by repetitive narrow pulses (ns–µs) with magnitudes in the kV range. The tip of the plasma plume is brought against the external wall of the chamber to ignite a plasma at reduced pressure.

3. Results and Discussion

For all the experiments described below, the pulsed power supply and its associated circuitry were all placed inside a Faraday cage located in a separate location, away from the chamber and the diagnostics equipment, and all cables were properly shielded. All spectroscopic and Langmuir probe measurements presented here reflect average values. No time- and space-resolved measurements are reported. In addition, in order to have a ground reference for the diagnostics circuitry, a thin copper ring was wrapped around one arm of the Pyrex chamber (externally) and connected to a hard ground. For all experiments, the jet was driven by the following parameters, unless otherwise mentioned: Voltage was 7 kV, pulse width was 1 µs, repetition rate was 7 kHz, and helium flow rate was 7 slm.

3.1. Measurement of the Deposited Charge

The electric field generated behind a dielectric barrier is a function of the amount of charges deposited on the outer/opposite surface of the dielectric. To measure the charges deposited on the surface of a dielectric surface by the plume of a plasma jet, the following circuit was used (Figure 4) [11]. The charge was calculated by integrating the current flowing through the readout resistor. The current through the resistor flowed via capacitive coupling, with the capacitance having the dielectric constant of the Pyrex glass, and the area was that of the copper plate electrode placed against the dielectric.

Figure 5 is a plot of the deposited charges versus time when the plasma jet is powered by 1 µs wide pulses of magnitude of 7 kV.

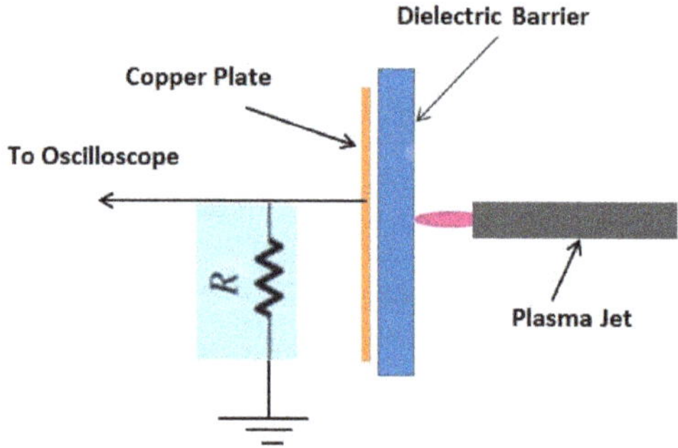

Figure 4. Circuit used to measure the amount of charge deposited on the surface of a dielectric. Copper plate was 1.5 inches × 1.5 inches thick. Dielectric plate was 2 inches × 2 inches and 0.6 inches thick. Resistance was 1 kΩ.

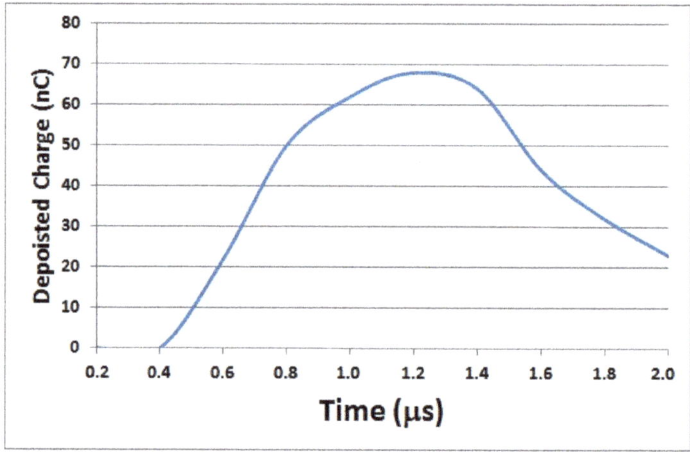

Figure 5. Deposited charge as a function of time for pulse 1 μs wide, magnitude of 7 kV, pulse frequency of 7 kHz, and helium flow rate of 7 slm.

The magnitude of the electric field generated by such charge accumulation was estimated to be in the 10–15 kV/cm range for a dielectric thickness of about 1.5 cm. This is in relative agreement with the spectroscopic measurements presented in the next section.

3.2. Measurement of the Transmitted Electric Field

The electric field was measured using the Stark splitting and shifting of the helium visible lines and their forbidden lines. The displacement of the Stark sublevels of He I 447.1 nm and its forbidden component was calculated. The wavelength separation ($\Delta\lambda_{\text{allowed-forbidden}}$) of the π components of allowed and forbidden lines ($m_{\text{upper}} = 0$ to $m_{\text{lower}} = 0$) was measured and its relationship to the electric field strength was used to calculate E using a polynomial relation [12,13] (see Equation (1)). Using this method, electric field strength up to 18 kV/cm was calculated [11]. Figure 6 shows the experimental setup while Figure 7 shows a result of such a measurement.

$$\Delta\lambda_{\text{allowed-forbidden}} = -1.6 \times 10^{-5} \times E^3 + 5.95 \times 10^{-4} \times E^2 + 2.5 \times 10^{-4} \times E + 0.15, \tag{1}$$

where E is expressed in kV/cm.

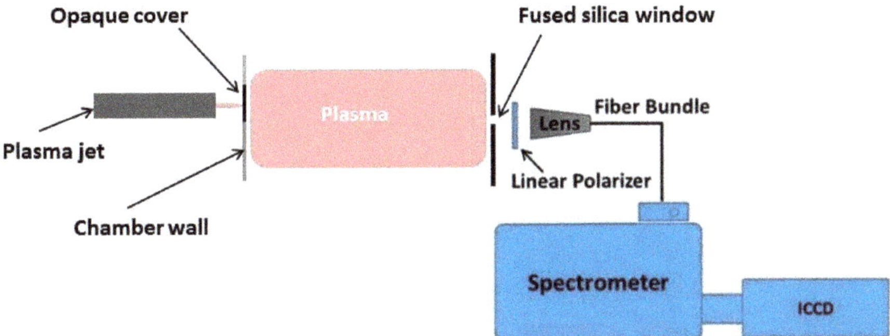

Figure 6. Sketch of the experimental setup to measure the electric field in the chamber. The electric field in the bulk of the plasma is assumed to be weak so the radiation collected to measure the axial electric field in the chamber gives a measure of the field close to the chamber wall (opposite side to where the jet impinges on the wall), where the field is high. The spectrometer was a 0.5 m Spectra-Pro-500i imaging spectrometer (Acton Research). The grating was 3600 g/mm, the slit width was 300 μm, and the spectral resolution was set at 0.02 nm. ICCD was a Dicam-Pro. The pressure inside the chamber was 1.5 Torr and the gas was helium. For each wavelength, more than a million ICCD images were integrated over a collection time of 100 ms.

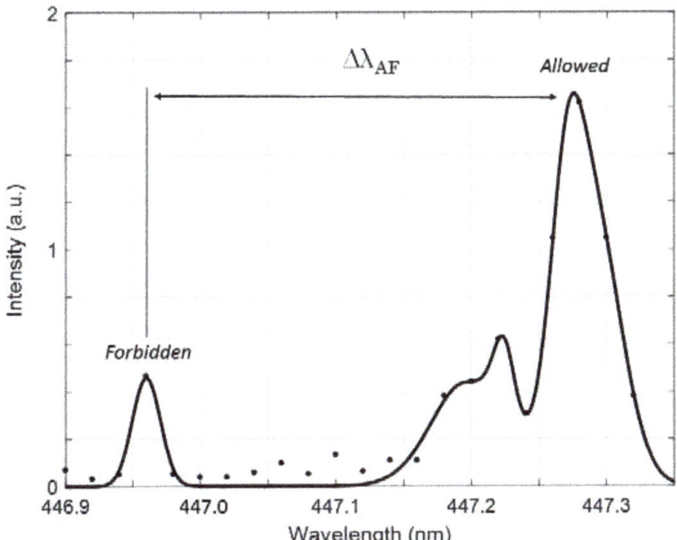

Figure 7. Electric field (EF) measurement using Stark splitting. An EF average strength of about 18 kV/cm was measured.

3.3. Measurement of the Electrons Density and Temperature

A Langmuir probe was used to measure the electrons density and temperature of the plasma discharge inside the chamber. It is important to note again that only time-averaged values were measured. The measurements presented here are preliminary and only meant to give an approximate idea on the order of magnitude of the density and temperature. The probe was made of a shielded 5 mm long metal needle with a surface area of 2.36 mm^2. A variable DC power supply was used to bias the probe. For an air plasma at a pressure of 400 mTorr, the average electron density was found to be around 1.6×10^{10} cm^{-3} and the average electron temperature was 2.2 eV. For a helium plasma, with a small admixture of air, at a pressure of 300 mTorr, the electron density was around 6.4×10^{10} cm^{-3} and the average temperature was 2.7 eV.

3.4. Dynamics of the Plasma Ignition Inside the Chamber

ICCD images of the expansion of the diffuse plasma inside the chamber, taken 20 nanoseconds apart, are shown in Figure 8 (using a Dicam Pro ICCD). The images show that, as the ionization wave front (i.e., plasma bullet) impinged on the outer surface of the chamber, a discharge was ignited behind the Pyrex glass wall. A glowing and expanding plasma advanced radially in all directions inside the chamber. The speed of expansion was estimated to be in the 10^6–10^7 cm/s range, consistent with that of an ionization wave in gases at low pressure.

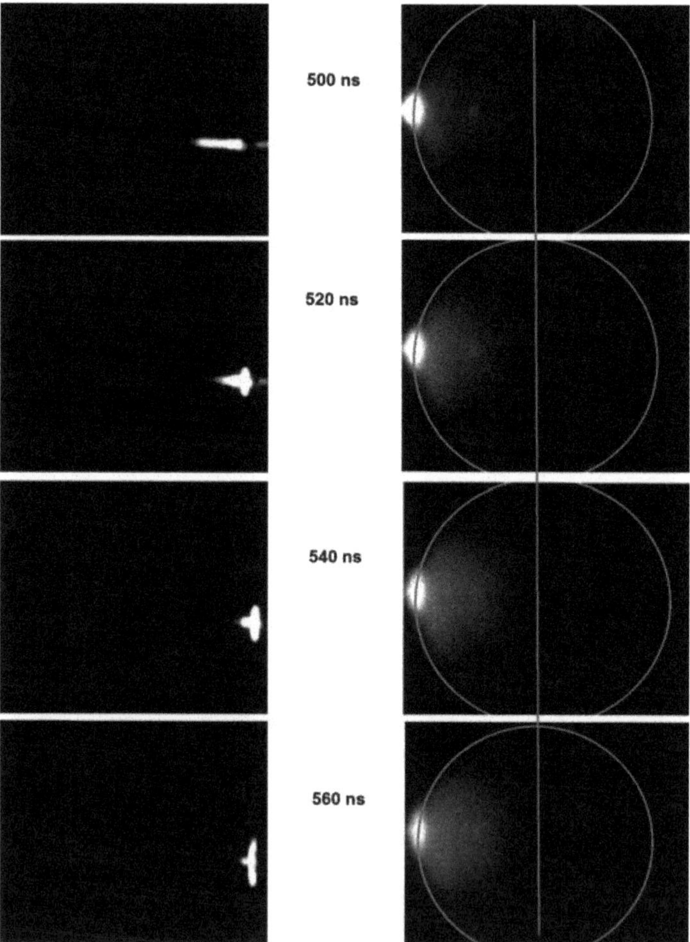

Figure 8. ICCD images of the plasma bullet (left panel) and of the expanding discharge inside the Pyrex tubing/chamber (right panel). The air pressure in the chamber was around 1 Torr. The contour of the chamber wall (grey circle) is added for illustrative purposes. The vertical line is added as a reference to better visualize the expansion of reduced pressure plasma. The images on the left panel show the plasma bullet arriving and spreading over the outer surface of the Pyrex glass wall.

4. Conclusions

Reduced pressure diffuse plasma can be generated inside a chamber having dielectric walls using an externally applied ionization wave emitted by a plasma jet. The chamber had no electrodes or any electrical connection. The dynamics of the diffuse plasma was elucidated using ICCD images, the electric field transmitted inside the chamber was measured using a spectroscopic technique, and the electron density and temperature were measured by a Langmuir probe. The advantage of the generation method discussed in this paper is that the plasma inside the chamber is free of metal contamination (since there are no electrodes). Such plasma can be very useful for material processing. Most low-pressure RF plasmas (capacitively or inductively coupled) require impedance matching when operated at high power. However, the new method described in this paper does not need an impedance matching module since the plasma jet is electrically independent of the reduced pressure plasma chamber. It only serves as the source of the ionization wave. Therefore, the plasma inside the chamber may be characterized as a "remotely" generated plasma.

Conflicts of Interest: The author declares no conflict of interest.

References

1. Begum, A.; Laroussi, M.; Pervez, M.R. Atmospheric pressure helium/air plasma jet: Breakdown processes and propagation phenomenon. *AIP Adv.* **2013**, *3*, 062117. [CrossRef]
2. Sobota, A.; Guaitella, O.; Garcia-Caurel, E. Experimentally obtained values of electric field of an atmospheric pressure plasma jet impinging on a dielectric surface. *J. Phys. D Appl. Phys.* **2013**, *46*, 37. [CrossRef]
3. Stretenovic, G.B.; Krstic, I.B.; Kovacevic, V.V.; Obradovic, A.M.; Kuraica, M.M. Spatio-temporally resolved electric field measurements in helium plasma jet. *J. Phys. D Appl. Phys.* **2014**, *47*, 10.
4. Robert, E.; Darny, T.; Dozias, S.; Iseni, S.; Pouvesle, J.-M. New insights on the propagation of pulsed atmospheric plasma streams: From single jet to multi jet arrays. *Phys. Plasmas* **2015**, *22*, 122007. [CrossRef]
5. Naidis, G.V. Modeling of Streamer Propagation in Atmospheric Pressure Helium Plasma Jets. *J. Phys. D Appl. Phys.* **2010**, *43*, 40. [CrossRef]
6. Lu, X.; Xiong, Q.; Xiong, Z.; Hu, J.; Zhou, F.; Gong, W.; Xian, Y.; Zou, C.; Tang, Z.; Jiang, Z.; et al. Propagation of an atmospheric pressure plasma plume. *J. Appl. Phys.* **2009**, *105*, 043304. [CrossRef]
7. Xiong, Z.; Robert, E.; Sarron, V.; Pouvesle, J.-M.; Kushner, M.J. Atmospheric-pressure plasma transfer across dielectric channels and tubes. *J. Phys. D Appl. Phys.* **2013**, *46*, 15. [CrossRef]
8. Guaitella, O.; Sobota, A. The impingement of a kHz helium atmospheric pressure plasma jet on a dielectric surface. *J. Phys. D Appl. Phys.* **2015**, *48*, 25. [CrossRef]
9. Laroussi, M.; Razavi, H. Indirect Generation of a Large Volume Diffuse Plasma by an Ionization Wave from a Plasma Jet. *IEEE Trans. Plasma Sci.* **2015**, *43*, 2226–2229. [CrossRef]
10. Lu, X.; Naidis, G.V.; Laroussi, M.; Ostrikov, K. Guided Ionization Waves: Theory and Experiments. *Phys. Rep.* **2014**, *540*, 123–166. [CrossRef]
11. Razavi, H.; Laroussi, M. Diagnostics of Diffuse Large Volume Plasma Generated by an External Ionization Waves. In Proceedings of the Gaseous Electronics Conference, Pittsburg, PA, USA, 6–10 November 2017; p. 42.
12. Kuraica, M.M.; Konjevic, N. Electric field measurement in the cathode fall region of a glow discharge in helium. *Appl. Phys. Lett.* **1997**, *70*, 1521. [CrossRef]
13. Lu, Y.; Wu, S.; Cheng, W.; Lu, X. Electric field measurements in an atmospheric-pressure microplasma jet using Stark polarization emission spectroscopy of helium atom. *Eur. Phys. J. Spec. Top.* **2017**, *226*, 2979–2989. [CrossRef]

 *plasma*

Editorial

Special Issue on Plasma Medicine

Mounir Laroussi [1,*], Michael Keidar [2] and Masaru Hori [3]

[1] Electrical & Computer Engineering Department, Old Dominion University, Norfolk, VA 23529, USA
[2] School of Engineering and Applied Science, George Washington University, Washington, DC 20052, USA; keidar@gwu.edu
[3] Institute of Innovation for Future Society, Nagoya University, Nagoya 464-8603, Japan; hori@nuee.nagoya-u.ac.jp
* Correspondence: mlarouss@odu.edu; Tel.: +757-683-6369

Received: 2 October 2018; Accepted: 15 October 2018; Published: 17 October 2018

Research on the applications of atmospheric pressure low temperature plasma (LTP) in biology and medicine started in the mid-1990s with experiments on the inactivation of bacteria on biotic and abiotic surfaces and in liquid media [1,2]. This was soon followed by investigations on the effects of LTP on mammalian cells [3–9]. The encouraging results obtained in these early works led to the consideration of LTP technology for new potential therapies in wound healing, dentistry, and cancer treatment [5–9]. By the end of the first decade of the 2000s, many LTP sources had been approved for medical use. These included the Rhytec Portrait® for use in dermatology (approved in 2008), the Bovie J-Plasma® (Clearwater, FL, USA), the Canady Helios Cold Plasma and Hybrid Plasma™ (Takoma Park, MD, USA) Scalpel, the Adtec MicroPLaSter® (Hounslow, UK, approved for clinical trials in 2008), the kINPen® (developed by INP, Greifswald, Germany, and medically certified as class IIa in 2013), and the PlasmaDerm® device (CINOGY GmbH, Duderstadt, Germany). In addition to potential applications in medicine here on Earth, LTP may prove to be a crucial technology for space medicine. In long-duration manned deep space missions, using LTP for decontamination and wound treatment, for example, would be a more suitable/applicable option than transporting and storing perishable chemical-based medication. In this context. LTP offers energy-based medical options that mostly require the availability of electrical power.

The effects of LTP on biological cells are believed to be mainly mediated by its reactive oxygen species (ROS) and reactive nitrogen species (RNS) [10,11]. These include hydroxyl, OH, atomic oxygen, O, singlet delta oxygen, $O_2(^1\Delta)$, superoxide, $O_2{}^-$, hydrogen peroxide, H_2O_2, and nitric oxide, NO. These species (radicals and non-radicals) can interact with cells membranes, enter the cells, and increase the intracellular ROS concentrations, which may lead to DNA damage and may compromise the integrity of other organelles and macromolecules [12–16]. ROS and RNS can also trigger cell signaling cascades, which can ultimately lead to cellular death pathways, such as apoptosis. Other plasma-generated agents that may play biological roles are charged particles and photons. In addition, LTP can exhibit large electric fields that are suspected to also play a role, such as in cellular electroporation, allowing large molecules to enter the cells.

This Special Issue contains eight papers discussing the latest results on the application of LTP to various cell lines and tissues. These papers discuss a variety of plasma medicine topics, including the treatment of ovarian cancer, triple-negative breast cancer, malignant solid tumors, new LTP devices, as well as a mini review and a paper describing atomic scale simulations on glucose uptake under LTP treatment.

To conclude, the guest editors would like to thank all the authors for their valuable contributions and the reviewers for their time and efforts.

References

1. Laroussi, M. Sterilization of Contaminated Matter with an Atmospheric Pressure Plasma. *IEEE Trans. Plasma Sci.* **1996**, *24*, 1188. [CrossRef]
2. Laroussi, M. Sterilization of Liquids Using Plasma Glow Discharge. U.S. Patent 5,876,663, 2 March 1999.
3. Shekhter, A.B.; Kabisov, R.K.; Pekshev, A.V.; Kozlov, N.; Perov, Y.L. Experimental and Clinical Validation of Plasmadynamic Therapy of Wounds with Nitric Oxide. *Bull. Exp. Biol. Med.* **1998**, *126*, 829. [CrossRef]
4. Stoffels, E.; Flikweert, A.J.; Stoffels, W.W.; Kroesen, G.M.W. Plasma Needle: A non-destructive Atmospheric Plasma Source for Fine Surface Treatment of Biomaterials. *Plasma Sources. Sci. Technol.* **2002**, *11*, 383. [CrossRef]
5. Laroussi, M.; Kong, M.; Morfill, G.; Stolz, W. *Plasma Medicine: Applications of Low Temperature Gas Plasma in Medicine and Biology*; Cambridge Univ. Press: Cambridge, UK, 2012; ISBN 978-1-107-00643-0.
6. Keidar, M.; Beilis, I.I. *Plasma Engineering: Application in Aerospace, Nanotechnology and Bio-Nanotechnology*; Elsevier: Oxford, UK, 2013; ISBN 978-0-123-85977-8.
7. Fridman, A.; Friedman, G. *Plasma Medicine*; Wiley: New York, NY, USA, 2013; ISBN 978-0-470-68970-7.
8. Metelmann, H.-R.; Von Woedtke, T.; Weltmann, K.-D. *Comprehensive Clinical Plasma Medicine*; Springer: Berlin, Germany, 2018; ISBN 978-3-319-67627-2.
9. Toyokuni, S.; Ikehara, Y.; Kikkawa, F.; Hori, M. *Plasma Medical Science*; Academic Press: Cambridge, MA, USA, 2018; ISBN 978-0-128-15004-7.
10. Graves, D. The emerging role of reactive oxygen and nitrogen species in redox biology and some implications for plasma applications to medicine and biology. *J. Phys. D* **2012**, *45*, 263001. [CrossRef]
11. Lu, X.; Naidis, G.V.; Laroussi, M.; Reuter, S.; Graves, D.B.; Ostrikov, K. Reactive Species in Non-equilibrium Atmospheric Pressure Plasma: Generation, Transport, and Biological Effects. *Phys. Rep.* **2016**, *630*, 1–84. [CrossRef]
12. Keidar, M.; Walk, R.; Shashurin, A.; Srinivasan, P.; Sandler, A.; Dasgupta, S.; Ravi, R.; Guerrero-Preston, R.; Trink, B. Cold Plasma Selectivity and the Possibility of a Paradigm Shift in Cancer Therapy. *Br. J. Cancer* **2011**, *105*, 1295. [CrossRef] [PubMed]
13. Schlegel, J.; Koritzer, J.; Boxhammer, V. Plasma in Cancer Treatment. *Clin. Plasma Med.* **2013**, *1*, 2. [CrossRef]
14. Utsumi, F.; Kjiyama, H.; Nakamura, K.; Tanaka, H.; Mizuno, M.; Ishikawa, K.; Kondo, H.; Kano, H.; Hori, M.; Kikkawa, F. Effect of Indirect Nonequilibrium Atmospheric Pressure Plasma on Anti-Proliferative Activity against Chronic Chemo-Resistant Ovarian Cancer Cells In Vitro and In Vivo. *PLoS ONE* **2013**, *8*, e81576. [CrossRef] [PubMed]
15. Laroussi, M.; Mohades, S.; Barekzi, N. Killing of Adherent and non-adherent Cancer Cells by the Plasma Pencil. *Biointerphases* **2015**, *10*, 029410. [CrossRef]
16. Laroussi, M. Effects of Low Temperature Plasmas on Proteins. *IEEE Trans. Radiat. Plasma Med. Sci.* **2018**, *2*, 229. [CrossRef]

 *plasma*

Review

Plasma Medicine: A Brief Introduction

Mounir Laroussi

Electrical & Computer Engineering Department, Old Dominion University, Norfolk, VA 23529, USA;
mlarouss@odu.edu; Tel.: +757-683-6369

Received: 28 January 2018; Accepted: 17 February 2018; Published: 19 February 2018

Abstract: This mini review is to introduce the readers of *Plasma* to the field of plasma medicine. This is a multidisciplinary field of research at the intersection of physics, engineering, biology and medicine. Plasma medicine is only about two decades old, but the research community active in this emerging field has grown tremendously in the last few years. Today, research is being conducted on a number of applications including wound healing and cancer treatment. Although a lot of knowledge has been created and our understanding of the fundamental mechanisms that play important roles in the interaction between low temperature plasma and biological cells and tissues has greatly expanded, much remains to be done to get a thorough and detailed picture of all the physical and biochemical processes that enter into play.

Keywords: low temperature plasma; plasma jet; cells; tissue; apoptosis; cancer; wound healing; reactive species

---

1. Introduction

In the mid-1990s, experiments were conducted that showed that low temperature atmospheric pressure plasmas (LTP) can be used to inactivate bacteria [1]. Based on these results, the Physics and Electronics Directorate of the US Air Force Office of Scientific Research (AFOSR) funded a proof of principle research program in 1997 and supported such research for a number of years. The results from this research program were widely disseminated in the literature, including in peer-reviewed journals and conference proceedings, therefore attracting the attention of the plasma physics community to new and emerging applications of low temperature plasma in biology and medicine [2–8]. The goals of the AFOSR program were to apply low temperature plasmas (LTP) to treat the wounds of injured soldiers and to sterilize/disinfect both biotic and abiotic surfaces. By the early 2000s, research expanded to include eukaryotic cells when small doses of LTP were found to enhance phagocytosis, accelerate the proliferation of fibroblasts, detach mammalian cells without causing necrosis, and under some conditions, lead to apoptosis [9,10].

The above-described groundbreaking research efforts showed that nonthermal plasma can gently interact with biological cells (prokaryotes and eukaryotes) to induce certain desired outcomes. These early achievements raised great interest and paved the way for many laboratories from around the world to investigate the biomedical applications of LTP and by the end of the first decade of the 2000s, a global scientific community was established around such research activities. The field is today known by the term plasma medicine, and in the last few years a number of extensive reviews and tutorials were published (see Refs [11–18] and references therein) as well as a few books [19–21].

Today, the field of plasma medicine encompasses several applications of low temperature plasmas in biology and medicine [22–53]. These include:

- Sterilization, disinfection, and decontamination,
- plasma-aided wound healing

- plasma dentistry
- cancer applications or "plasma oncology,"
- plasma pharmacology,
- plasma treatment of implants for biocompatibility.

In the late 2000s, several LTP sources were approved for cosmetic and medical use. Examples are: in 2008 the US FDA approved the Rhytec Portrait® (plasma jet) for use in dermatology. Also in the US other plasma devices are in use today for various medical applications, such as the Bovie J-Plasma® and the Canady Helios Cold Plasma and Hybrid Plasma™ Scalpel. In Germany, the medical device certification class IIa was given to the kINPen® (plasma jet) in 2013, and the PlasmaDerm® device (CINOGY GmbH) was also approved. Figure 1 is a timeline graph showing the major milestones in the development of the field of low temperature plasma medicine.

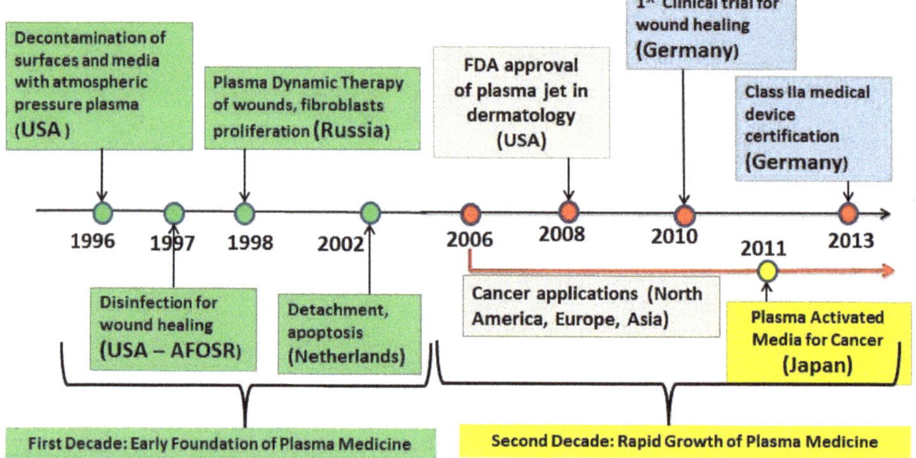

Figure 1. Timeline showing some major milestones of the new field of the biomedical applications of low temperature atmospheric pressure plasma. This timeline does not show the case of thermal (hot) plasmas, which were used for many decades in medical applications requiring heat, such as cauterization and blood coagulation.

2. LTP Takes on Hygiene and Medical Challenges

As can be seen from Figure 1 the biomedical applications of LTP started with experiments on the inactivation of bacteria on biotic and abiotic surfaces and media. Bacterial contamination proved to pose severe challenges for some industries and in the healthcare arena. The industrial challenges are mainly around the problem of food contamination and sterilization of food packaging. Several well-publicized food poisoning incidents (EHEC, *Listeria*, *Salmonella*) pointed out to consumers that the present methods employed by the food industry may not be adequate to insure food safety. The healthcare challenges are linked to nosocomial infections caused by antibiotic resistant strains of bacteria, such as Methicillin-resistant *Staphylococcus aureus* (MRSA) and *Clostridium difficile* (C-diff). Every year in the US, hospital acquired infections (HAI) kill thousands of patients with compromised immune systems. HAI are caused by inadequate sterilization/decontamination of instruments, surfaces, clothing, bedsheets, and personnel (nurses and doctors). In most cases, contamination by strains of bacteria resistant to the best antibiotic medications available today is the cause of HAI. LTP is therefore considered as a novel technology that can be successfully applied to help solve some of the challenges described above.

The most recent application presently receiving much attention is the use of LTP to destroy cancer cells and tumors in a selective manner [38–58]. Starting around the mid-2000s several investigators reported experiments showing that low temperature plasmas (LTP) can destroy cancerous cells in vitro. This was followed by some in vivo work showing that LTP can reduce the size of cancer tumors in animal models. The in vitro work covered a host of cancerous cell lines, which included glioblastoma, melanoma, papilloma, carcinoma, colorectal cancer, ovarian cancer, prostate cancer cells, squamous cell carcinoma, leukemia, and lung cancer. The in vivo (animal model) work can be found in [38,44,45,53].

In addition to direct plasma applications to cancer cells and tissues, investigators reported that plasma-activated media (PAM) can also be used to destroy cancer cells [38,50,54–58]. Plasma-activated medium is produced by exposing a biological liquid medium to LTP for a length of time (minutes). In this case, the plasma-generated reactive species interact with the contents of the medium and generate solvated long-lived reactive species in the liquid, such as hydrogen peroxide, H_2O_2, nitrite, NO_2^-, nitrate, NO_3^-, peroxynitrite. $ONOO^-$, and organic radicals. These molecules subsequently react with the cells and tissues causing various biological outcomes.

3. Mechanisms of Biological Action of LTP: Brief Summary

Investigators reported that the effects of LTP on biological cells (prokaryotes and eukaryotes) are mediated by reactive oxygen and nitrogen species (RONS) [11,12,59–66]. These species include hydroxyl, OH, atomic oxygen, O, singlet delta oxygen, $O_2(^1\Delta)$, superoxide, O_2^-, hydrogen peroxide, H_2O_2, and nitric oxide, NO. For example, the hydroxyl radical is known to cause the peroxidation of unsaturated fatty acids, which make up the lipids constituting the cell membrane. The biological effects of hydrogen peroxide are mediated by its strong oxidative properties affecting lipids, proteins, and DNA (single and possibly double-strand breaks). Nitric oxide, which acts as an intracellular messenger and regulator in biological functions, is known to affect the regulation of immune deficiencies, cell proliferation, induction of phagocytosis, regulation of collagen synthesis, and angiogenesis.

In cancer cells, the mechanisms of action of LTP are suspected to be related to an increase of intracellular reactive oxygen species (ROS), which can lead to cell cycle arrest at the S-phase, DNA double-strand breaks, and induction of apoptosis. Research by various groups showed that RONS generated by LTP react with cell membranes and can even penetrate the cells and induce subsequent reactions within the cells that can trigger cell-signaling cascades, which can ultimately lead to apoptosis in cancer cells [56–66]. In addition, investigators have shown that plasma-generated RONS can indeed penetrate biological tissues up to depths of more than 1 mm and therefore interact not only with the cells on the surface but with those underneath [67–72].

LTP delivers not only reactive species but it also can exhibit large enough electric fields [73–77]. The magnitudes of these electric fields are several kV/cm and they are suspected to play a role, such as in cellular electroporation, which may allow large molecules to enter the cells.

4. Two LTP Sources for Biomedical Applications: Brief Description

The main LTP devices used in plasma medicine research are the dielectric barrier discharge (DBD) and nonequilibrium atmospheric pressure plasma jets (N-APPJ). In fact, the DBD was the device used in the first experiments on the inactivation of bacteria [1]. The DBD uses plate electrodes covered by a dielectric (such as glass). The plasma is generated in the gap separating the electrodes by the application of high sinusoidal voltages in the kHz frequency range. Gases such helium with admixtures of oxygen or air are usually used. For more information on the working of the DBD see references [61,78,79]. Figure 2 shows a schematic of the DBD and a photograph of a diffuse plasma at atmospheric pressure generated by a DBD.

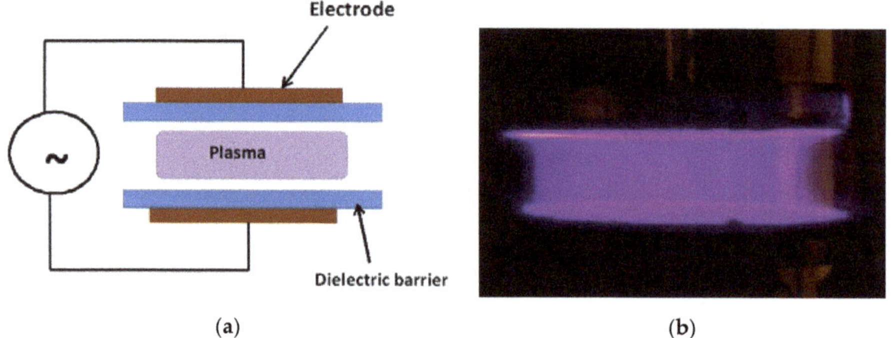

Figure 2. Schematic (**a**) and a photograph (**b**) of an atmospheric pressure diffuse plasma generated by a dielectric barrier discharge (DBD). The discharge in the photo on the right is driven by kHz sinusoidal high voltage and the gas is helium with a small admixture of air. Photo taken at the author's laboratory.

Nonequilibrium atmospheric pressure plasma jets (N-APPJs) produce plasma plumes that propagate away from the confinement of electrodes and into the ambient air. The reactive species generated by the plasma can therefore safely and conveniently be transported to a target at a remote location and away from the main plasma generation area. This characteristic made N-APPJs very attractive tools for applications in biology and medicine [60,80–82]. Various power driving methods that include pulsed DC, RF, and microwave power have been used. In addition, various electrode configurations ranging from single electrode, to two-ring electrodes wrapped around the outside wall of a cylindrical dielectric body, to two-ring electrodes attached to centrally perforated dielectric disks have been used. Figure 3 shows photographs of two N-APPJs, the plasma pencil and the kINPen, which have been used extensively in plasma medicine research.

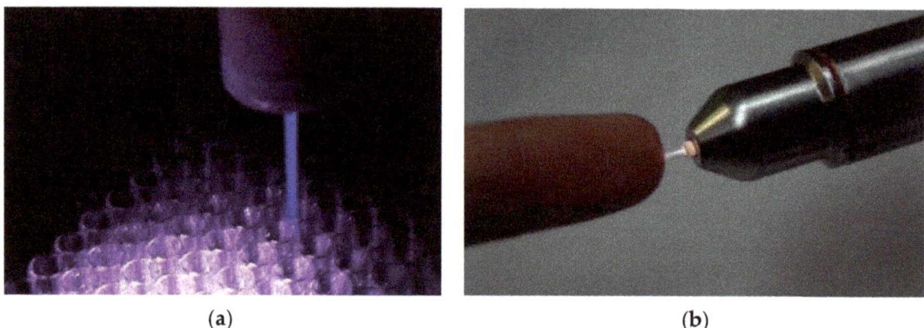

Figure 3. Photographs of two plasma jets that have been used in various biomedical applications. (**a**) is the plasma pencil (ODU, Norfolk, VA, USA), and (**b**) is the kINPen (INP, Greifswald, Germany).

The plasma plumes emitted by N-APPJs turned out to be made of small plasma packets traveling at very high velocities (tens of km/s). These plasma packets came to be known as "plasma bullets" and they were independently first reported in the mid-2000s by Teschke et al and by Lu and Laroussi [83,84]. Lu and Laroussi used nanosecond-pulsed DC power while Teschke et al used RF power. The plasma bullets were subsequently researched extensively, both experimentally and by modeling, by various investigators [85–91]. Today there is agreement that the plasma bullets are guided ionization waves. To learn about these guided ionization waves in greater detail, the reader is referred to [92].

5. Two Biomedical Applications of LTP

To illustrate the effects of LTP on biological targets, two applications are shown here. The first concerns the bactericidal property of LTP and the second shows the effects of direct plasma exposure as well as plasma activated media on cancerous and healthy epithelial cells. The results presented below are based on the use of the plasma pencil described earlier. The results shown were obtained by the application of the LTP plume generated by the plasma pencil on a bacterial lawn seeded on the surface of a Petri dish (see Figure 4). The bacterium used was *Acinetobacter calcoaceticus*, a gram-negative soil bacterium also found in the tiger mosquito, which is known to be a transmission vector of yellow and dengue fevers. Figure 5 shows zones of inactivation (dark circular areas) around the center of the dish where the plasma plume was applied. The photo to the left is for an initial bacteria concentration of 10^9/mL, while that on the right is for an initial concentration of 10^7/mL. It is clear that the killing effects are more extended and pronounced for the lower initial concentration. For more information on the dependence of inactivation on the plasma exposure time and on the type of bacteria, the reader is referred to [93].

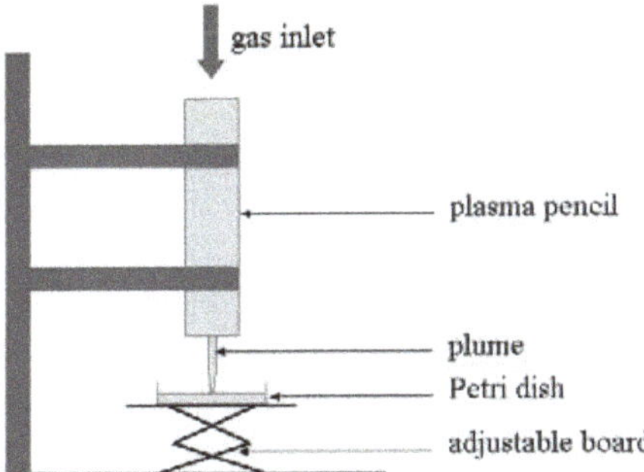

Figure 4. Experimental setup for the bacterial inactivation experiments.

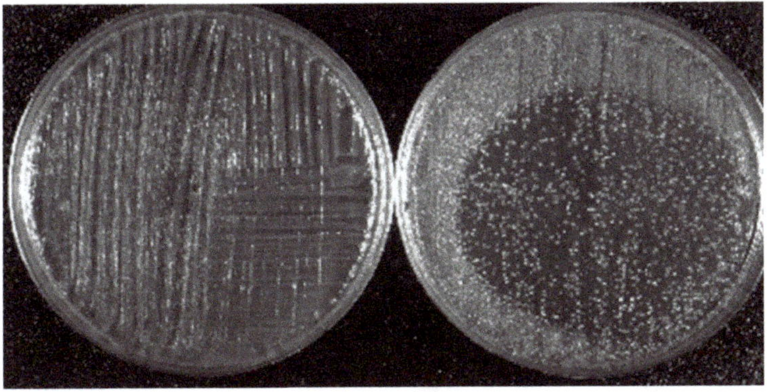

Figure 5. Killing property of LTP: Dependence of the killing efficacy on the initial bacteria concentration. Left picture is for 10^9/mL and right picture is for 10^7/mL. Bacterium is *A. calcoaceticus*. LTP source is the plasma pencil operated with helium as a carrier gas [93].

Figure 6 shows the effects of direct application of LTP on suspensions of cancerous cells. The cancer cell line used was a squamous cell carcinoma of the bladder (SCaBER, ATCC HTB-3™) originally obtained from a human bladder. After LTP exposure and proper incubation process (37 °C under 5% CO_2 atmosphere), Trypan-blue exclusion assay was used to count the number of live and dead cells. For details of the experimental protocol please refer to [40]. The counts immediately after LTP treatment (at 0 h) revealed no dead cells, which suggested there were no immediate physical effects. However, the viability of cells reduced to around 50% at 24 h after a 2-min LTP treatment. As seen in Figure 6, higher plasma exposure times result in more cells killed (5-min plasma treatment results in 75% of loss of viability at 24 h post-treatment) [40]. These results indicate that LTP does not apply immediate brute physical force on the cells, but its effects require longer biological times to show. This is an indication that plasma agents, such as reactive species and electric fields, interact with the cells and induce reactions and/or trigger biochemical pathways that ultimately result in the death of the cancer cells hours later.

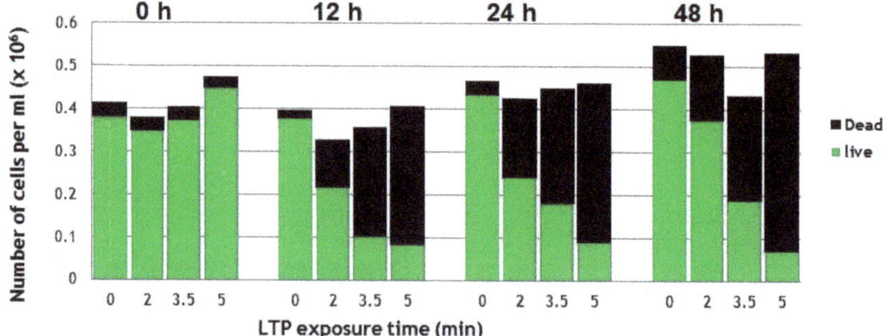

Figure 6. Viability of SCaBER cells in media treated directly by the LTP plume of the plasma pencil reveal dead (black bars on top) and live (green bars) cells. The viability was monitored at 0, 12, 24 and 48h post-LTP treatment [40].

Figure 7 shows the selective effect of LTP when it comes to destroying cancer cells versus healthy cells in vitro. The viability results shown in the figure below were obtained using plasma activated media (PAM), which was created by exposing biological liquid media to the plasma pencil for certain lengths of time. The cancerous cell line used was SCaBER and the healthy/normal cells were MDCK (Madin-Darby canine kidney) cells from normal epithelial tissue of a dog kidney. The media used to make PAM were MEM (minimum essential media) for SCaBER and Eagle Minimum Essential Media (EMEM) for MDCK. Figure 7 shows the results [57].

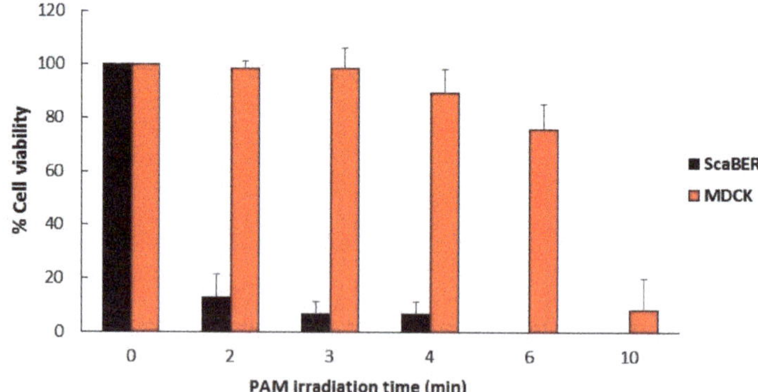

Figure 7. Viability in percent of SCaBER (cancerous) and MDCK cells (noncancerous) treated by PAM for various lengths of time. Viability was assessed after 12 hours incubation with PAM using MTS assay and Trypan-blue exclusion assay [57].

Figure 7 shows that PAM created using longer exposures to LTP has increasing killing effects on SCaBER cancer cells, reducing their viability to below 10% for irradiation times longer than 2 min. However, normal MDCK cells were able to withstand exposure to PAM for 3 min. This illustrates the selectivity of PAM in killing cancer cells while sparing healthy cells. But for PAM created with longer exposures to LTP (6 minutes and more) extensive killing of MDCK cells was obtained. This illustrates that the plasma dose is an important factor to take into consideration for optimal outcomes.

6. Penetration of RONS in Tissues

One of the key questions in plasma medicine is the following: Do the RONS generated by LTP only interact and affect cells on the surface of a tissue (or tumor) or do they penetrate the tissue and affect cells in deeper layers? Experimental evidence has shown that LTP does indeed affect cells underneath the tissue surface but what remains unclear is how. One possible explanation is what is referred to as the "bystander effect," which implies that there are chemical signals sent by the cells on the surface (in contact with plasma) to cells in the layer below [41]. These signals would trigger reactions similar to those occurring at the cells on the surface, including the onset of apoptosis. However, and to the best of this author's knowledge, there has been no experimental proof this occurs when LTP interacts with tissues. So, the possibility is there, but reliable data that can be replicated needs to emerge first. Therefore, in this section, only experiments that reported qualitatively and/or quantitatively on the penetration of RONS are presented.

In order to qualitatively and quantitatively elucidate RONS penetration into tissues, investigators used various in vitro models. Oh et al. investigated the penetration of RONS using a model made of an agarose film covering a volume of deionized water contained in a quartz cuvette [67]. They found that RONS kept being delivered from the agarose film to deionized water underneath it for up to 25 min after the plasma was removed. To study the delivery of reactive oxygen species (ROS) into cells, Hong et al. used a model comprising phospholipids vesicles encapsulated within a gelatin matrix and equipped with reactive oxygen species (ROS) reporter [68]. They found that ROS were delivered to the cells without rupturing the membranes of the vesicles. To simulate biological tissue, Szili et al. used gelatin gel, a derivative of collagen, and reported on the penetration behavior of H_2O_2 through a 1.5 mm thickness gelatin film [69]. The same authors also investigated the effects on DNA in synthetic tissue fluids, tissue, and cells [94].

Tissue models are useful and provide preliminary data regarding the penetration of RONS through biological targets. However, to simulate more realistic conditions, Duan et al. used slices of pig muscle tissue of different thicknesses placed on top of a PBS solution [72]. Figure 8 shows the experimental setup. A plasma jet operated with a helium/oxygen mixture was used. To ignite the plasma sinusoidal high voltages at a frequency of 1 kHz were employed. The plasma treatment times were 0, 5, 10, and 15 min.

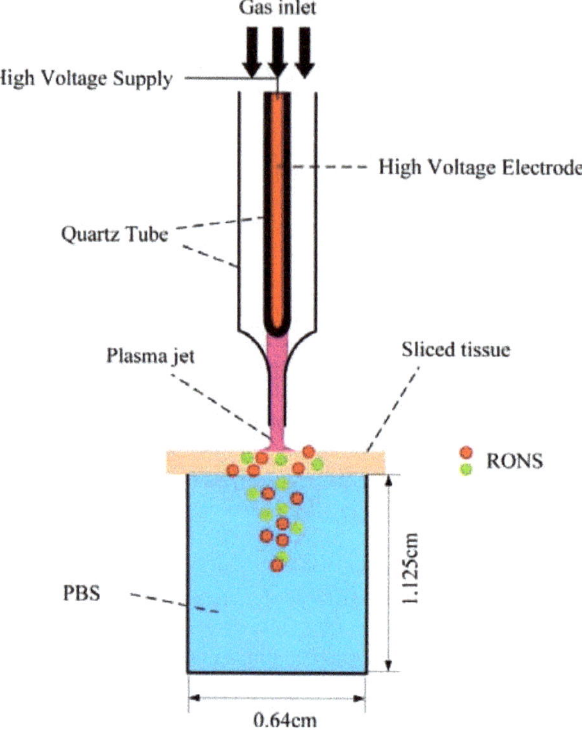

Figure 8. Experimental setup using pig muscle tissue [72]. Reproduced from Duan, J.; Lu, X.; and He, G. *Phys. Plasmas* **2017**, *24*, 073506, with the permission of AIP Publishing.

The concentrations of H_2O_2, OH, and that of the total of (NO_2^- + NO_3^-) were measured for different thicknesses of the tissue slice. A comparison of these concentrations when no tissue was used and when a tissue was placed on top of the solution showed that the concentrations of O_3, OH, and H_2O_2 were mostly consumed by the tissue and could not pass through 500-μm or greater tissue thickness. However, more than 80% of the (NO_2^- + NO_3^-) penetrated a 500-μm-thick tissue slice. Figure 9 shows the measured concentrations of (NO_2^- + NO_3^-) as a function of tissue thickness and for three plasma treatment times (5, 10, and 15 min).

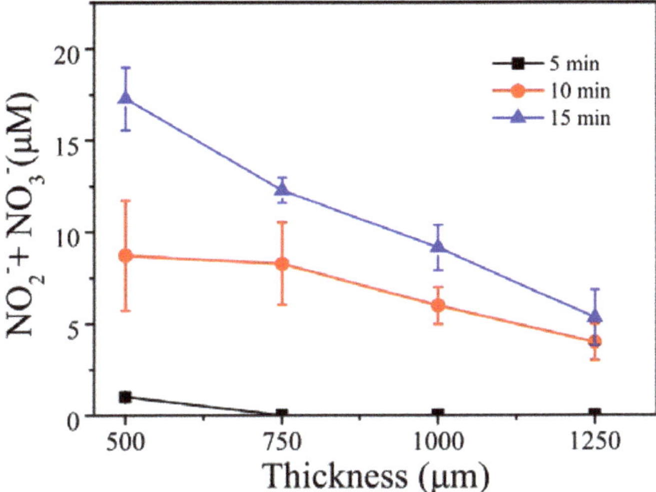

Figure 9. Total nitrite and nitrate concentration versus tissue thickness for three plasma exposure times [72]. Reproduced from Duan, J.; Lu, X.; and He, G. *Phys. Plasmas* **2017**, *24*, 073506, with the permission of AIP Publishing.

Figure 9 shows that the concentrations of the nitrogen reactive species, RNS, decrease with the tissue thickness, but increase with the plasma treatment time. The concentration of ($NO_2^- + NO_3^-$) for the 500-μm tissue thickness was comparable to the concentration when no tissue was placed on top of the PBS solution. This means that (RNS) were able to penetrate the tissue slice. This was not the case for ROS, which were absorbed by the tissue, unlike the case when a gelatin model (not real tissue) was used. For that model, ROS were able to penetrate the gelatin film.

The above examples illustrate that RONS do not simply react with the surface of tissues but can indeed penetrate relatively deeply. However, in more realistic conditions using actual tissue, it was shown that not all RONS can cross the same thickness. Some can be absorbed within a few tens of micrometers by the tissue, while others can penetrate up to 1.5 mm below the surface. Of course, the above results may not completely reflect what would happen under in vivo conditions. Such experiments need to be conducted and compared to results obtained for in vitro models and to those obtained under ex vivo conditions [95].

7. Conclusions

To conclude this brief introduction of the field of plasma medicine, it is safe to say that the biomedical applications of low temperature plasma have opened up an entirely new multidisciplinary field of research requiring close collaboration between physicists, engineers, biologists, biochemists, and medical experts. This multidisciplinary field started in mid-1990s with seminal experiments on the inactivation of bacteria by low temperature atmospheric pressure plasma generated by a dielectric barrier discharge and slowly expanded to include investigations on eukaryotic cells. Applications in dermatology, wound healing, dentistry, and cancer have led to various scientific advances and to the idea that LTP can be a technology upon which various innovative medical therapies can be developed to overcome present healthcare challenges. However, a lot remains to be done in order to fully understand the mechanisms of action of LTP against biological cells and tissues, both in vitro and in vivo. There is strong indication that LTP acts selectively on cancer cells and tumors and can penetrate deep below the surface, but much more work, including extensive clinical trials, is needed before LTP can be considered a safe technology ready for use in hospitals to treat chronic wounds, cancer lesions and tumors, and other ailments.

Conflicts of Interest: The author declares no conflicts of interest.

References

1. Laroussi, M. Sterilization of Contaminated Matter with an Atmospheric Pressure Plasma. *IEEE Trans. Plasma Sci.* **1996**, *24*, 1188. [CrossRef]
2. Garate, E.; Evans, K.; Gornostaeva, O.; Alexeff, I.; Kang, W.; Rader, M.; Wood, T. Atmospheric plasma induced sterilization and chemical neutralization. In Proceedings of the IEEE International Conference on Plasma Science, Raleigh, NC, USA, 1–4 June 1998. [CrossRef]
3. Laroussi, M.; Sayler, G.; Galscock, B.; McCurdy, B.; Pearce, M.; Bright, N.; Malott, C. Images of biological samples undergoing sterilization by a glow discharge at atmospheric pressure. *IEEE Trans. Plasma Sci.* **1999**, *27*, 34. [CrossRef]
4. Hermann, H.W.; Henins, I.; Park, J.; Selwyn, G.S. Decontamination of chemical and biological warfare (CBW) agents using an atmospheric pressure plasma jet (APPJ). *Phys. Plasmas* **1999**, *6*, 2284. [CrossRef]
5. Birmingham, J.G.; Hammerstrom, D.J. Bacterial decontamination using ambient pressure nonthermal discharges. *IEEE Trans. Plasma Sci.* **2000**, *28*, 51. [CrossRef]
6. Laroussi, M.; Alexeff, I.; Kang, W. Biological Decontamination by Non-thermal Plasma. *IEEE Trans. Plasma Sci.* **2000**, *28*, 184. [CrossRef]
7. Montie, T.C.; Kelly-Wintenberg, K.; Roth, J.R. An overview of research using the one atmosphere uniform glow discharge plasma (OAUGDP) for sterilization of surfaces and materials. *IEEE Trans. Plasma Sci.* **2000**, *28*, 41. [CrossRef]
8. Laroussi, M.; Richardson, J.P.; Dobbs, F.C. Effects of Non-Equilibrium Atmospheric Pressure Plasmas on the Heterotrophic Pathways of Bacteria and on their Cell Morphology. *Appl. Phys. Lett.* **2002**, *81*, 772. [CrossRef]
9. Shekhter, A.B.; Kabisov, R.K.; Pekshev, A.V.; Kozlov, N.P.; Perov, Y.L. Experimental and Clinical Validation of Plasmadynamic Therapy of Wounds with Nitric Oxide. *Bull. Exp. Biol. Med.* **1998**, *126*, 829. [CrossRef]
10. Stoffels, E.; Flikweert, A.J.; Stoffels, W.W.; Kroesen, G.M.W. Plasma Needle: A non-destructive Atmospheric Plasma Source for Fine Surface Treatment of Biomaterials. *Plasma Sources. Sci. Technol.* **2002**, *11*, 383. [CrossRef]
11. Fridman, G.; Friedman, G.; Gutsol, A.; Shekhter, A.B.; Vasilets, V.N.; Fridman, A. Applied plasma medicine. *Plasma Process. Polym.* **2008**, *5*, 503. [CrossRef]
12. Laroussi, M. Low Temperature Plasmas for Medicine. *IEEE Trans. Plasma Sci.* **2009**, *37*, 714. [CrossRef]
13. Barekzi, N.; Laroussi, M. Effects of Low Temperature Plasmas on Cancer Cells. *Plasma Process. Polym.* **2013**, *10*, 1039. [CrossRef]
14. Keidar, M.; Shashurin, A.; Volotskova, O.; Stepp, M.A.; Srinivasan, P.; Sandler, A.; Trink, B. Cold atmospheric plasma in cancer therapy. *Phys. Plasmas* **2013**, *20*, 057101. [CrossRef]
15. Laroussi, M.; Mohades, S.; Barekzi, N. Killing of Adherent and non-adherent Cancer Cells by the Plasma Pencil. *Biointerphases* **2015**, *10*, 029410. [CrossRef]
16. Laroussi, M. Non-Thermal Decontamination of Biological Media by Atmospheric Pressure Plasmas: Review, Analysis, and Prospects. *IEEE Trans. Plasma Sci.* **2002**, *30*, 1409. [CrossRef]
17. Laroussi, M. Low Temperature Plasma-Based Sterilization: Overview and State-of-the-Art. *Plasma Proc. Polym.* **2005**, *2*, 391. [CrossRef]
18. Von Woedtke, T.; Reuter, S.; Masur, K.; Weltmann, K.-D. Plasma for Medicine. *Phys. Rep.* **2013**, *530*, 291. [CrossRef]
19. Laroussi, M.; Kong, M.; Morfill, G.; Stolz, W. *Plasma Medicine: Applications of Low Temperature Gas Plasmas in Medicine and Biology*; Cambridge University Press: Cambridge, UK, 2012; ISBN 978-1-107-00643-0.
20. Fridman, A.; Friedman, G. *Plasma Medicine*; Wiley: New York, NY, USA, 2013; ISBN 978-0-470-68970-7.
21. Keidar, M.; Beilis, I.I. *Plasma Engineering: Applications from Aerospace to Bio and Nanotechnology*; Academic Press: London, UK, 2013; ISBN 978-0-12-385977-8.

22. Laroussi, M. Interaction of Low Temperature Plasma with Prokaryotic and Eukaryotic cells. In Proceedings of the 61st Gaseous Electronics Conference, Dallas, TX, USA, 13–17 October 2008; American Physical Society: Ridge, NY, USA; p. 29.

23. Isbary, G.; Morfill, G.; Schmidt, H.U.; Georgi, M.; Ramrath, K.; Heinlin, J. A first prospective randomized controlled trial to decrease bacterial load using cold atmospheric argon plasma on chronic wounds in patients. *Br. J. Dermatol.* **2010**, *163*, 78. [CrossRef] [PubMed]

24. Morris, A.D.; McCombs, G.B.; Akan, T.; Hynes, W.; Laroussi, M.; Tolle, S.L. Cold Plasma Technology: Bactericidal Effects on *Geobacillus Stearothermophilus* and *Bacillus Cereus* Microorganisms. *J. Dental Hygiene* **2009**, *83*, 55.

25. Claiborne, D.; McCombs, G.B.; Lemaster, M.; Akman, M.A.; Laroussi, M. Low Temperature Atmospheric Pressure Plasma Enhanced Tooth Whitening: The Next Generation Technology. *Int. J. Dent. Hygiene* **2013**. [CrossRef] [PubMed]

26. Laroussi, M.; Barekzi, N. Effects of Low Temperature Plasma on Two Eukaryotic Cell Lines: Epithelial Cells and Prostate Cancer Cells. In Proceedings of the 31st ICPIG, Granada, Spain, 14–19 July 2013.

27. Barekzi, N.; Laroussi, M. Fibropblasts Cell Morphology Altered by Low Temperature Atmospheric Pressure Plasma. *IEEE Trans. Plasma Sci.* **2014**, *42*, 2738. [CrossRef]

28. Laroussi, M.; Karakas, E.; Hynes, W. Influence of Cell Type, Initial Concentration, and Medium on the Inactivation Efficiency of Low Temperature Plasma. *IEEE Trans. Plasma Sci.* **2011**, *39*, 2960. [CrossRef]

29. Fridman, G.; Brooks, A.; Galasubramanian, M.; Fridman, A.; Gutsol, A.; Vasilets, V.; Ayan, H.; Friedman, G. Comparison of direct and indirect effects of non-thermal atmospheric-pressure plasma on bacteria. *Plasma Process. Polym.* **2007**, *4*, 370. [CrossRef]

30. Shashurin, A.; Keidar, M.; Bronnikov, S.; Jurjus, R.A.; Stepp, M.A. Living tissue under treatment of cold plasma atmospheric jet. *Appl. Phys. Lett.* **2008**, *93*, 181501. [CrossRef]

31. Laroussi, M.; VanWay, L.; Mohades, S.; Barekzi, N. Images of SCaBER Cells Treated by Low Temperature Plasma. *IEEE Trans. Plasma Sci.* **2014**, *42*, 2468. [CrossRef]

32. Xiong, Z.; Cao, Y.; Lu, X.; Du, T. Plasmas in tooth root canal. *IEEE Trans. Plasma Sci.* **2011**, *39*, 2968. [CrossRef]

33. Zimmermann, J.L.; Shimizu, T.; Boxhammer, V.; Morfill, G.E. Disinfection through different textiles using low-temperature atmospheric pressure plasma. *Plasma Process. Polym.* **2012**, *9*, 792. [CrossRef]

34. Babaeva, N.; Kushner, M.J. Reactive fluxes delivered by dielectric barrier discharge filaments to slightly wounded skin. *J. Phys. D: Appl. Phys.* **2013**, *46*, 025401. [CrossRef]

35. Weltmann, K.-D.; Kindel, E.; Brandenburg, R.; Meyer, C.; Bussiahn, C.; Wilke, C.; von Woedtke, T. Atmospheric Pressure Plasma Jet for Medical Therapy: Plasma Parameters and Risk Estimation. *Contrib. Plasma Plasma Phys.* **2009**, *49*, 631. [CrossRef]

36. Ehlbeck, J.; Schnabel, U.; Polak, M.; Winter, J.; von Woedtke, T.; Brandenburg, R.; von dem Hagen, T.; Weltmann, K.-D. Low temperature atmospheric pressure plasma sources for microbial decontamination. *J. Phys. D Appl. Phys.* **2011**, *44*, 013002. [CrossRef]

37. McCombs, G.B.; Darby, M.; Laroussi, M. "Dental Applications". In *Plasma Medicine: Applications of Low Temperature Gas Plasmas in Medicine and Biology*; Laroussi, M., Kong, M., Morfill, G., Stolz, W., Eds.; Cambridge University Press: Cambridge, UK, 2012.

38. Utsumi, F.; Kjiyama, H.; Nakamura, K.; Tanaka, H.; Mizuno, M.; Ishikawa, K.; Kondo, H.; Kano, H.; Hori, M.; Kikkawa, F. Effect of Indirect Nonequilibrium Atmospheric Pressure Plasma on Anti-Proliferative Activity against Chronic Chemo-Resistant Ovarian Cancer Cells In Vitro and In Vivo. *PLoS ONE* **2013**, *8*, e81576. [CrossRef] [PubMed]

39. Tanaka, H.; Mizuno, M.; Ishikawa, K.; Takeda, K.; Nakamura, K.; Utsumi, F.; Kajiyama, H.; Kano, H.; Okazaki, Y.; Toyokuni, S.; et al. Plasma Medical Science for Cancer Therapy: Toward Cancer Therapy Using Nonthermal Atmospheric Pressure Plasma. *IEEE Trans. Plasma Sci.* **2014**, *42*, 3760. [CrossRef]

40. Mohades, S.; Barekzi, N.; Laroussi, M. Efficacy of Low Temperature Plasma against SCaBER Cancer Cells. *Plasma Process. Polym.* **2014**, *11*, 1150. [CrossRef]

41. Laroussi, M. From Killing Bacteria to Destroying Cancer Cells: Twenty Years of Plasma Medicine. *Plasma Process. Polym.* **2014**, *11*, 1138. [CrossRef]

42. Köritzer, J.; Boxhammer, V.; Schäfer, A.; Shimizu, T.; Klämpfl, T.G.; Li, Y.-F.; Welz, C.; Schwenk-Zieger, F.; Morfill, G.E.; Zimmermann, J.L.; et al. Restoration of sensitivity in chemo-resistant glioma cells by cold atmospheric plasma. *PLos ONE* **2013**, *8*, e64498.

43. Schlegel, J.; Koritzer, J.; Boxhammer, V. Plasma in Cancer Treatment. *Clin. Plasma Med.* **2013**, *1*, 2. [CrossRef]
44. Vandamme, M.; Robert, E.; Pesnele, S.; Barbosa, E.; Dozias, S.; Sobilo, J.; Lerondel, S.; Le Pape, A.; Pouvesle, J.-M. Antitumor Effects of Plasma Treatment on U87 Glioma Xenografts: Preliminary Results. *Plasma Process. Polym.* **2010**, *7*, 264. [CrossRef]
45. Keidar, M.; Walk, R.; Shashurin, A.; Srinivasan, P.; Sandler, A.; Dasgupta, S.; Ravi, R.; Guerrero-Preston, R.; Trink, B. Cold Plasma Selectivity and the Possibility of a Paradigm Shift in Cancer Therapy. *Br. J. Cancer.* **2011**, *105*, 1295. [CrossRef] [PubMed]
46. Laroussi, M.; Keidar, M. Plasma & Cancer. *Plasma Process. Polym.* **2014**, *11*, 1118.
47. Fridman, G.; Shereshevsky, A.; Jost, M.M.; Brooks, A.D.; Fridman, A.; Gutsol, A.; Vasilets, V.; Friedman, G. Floating electrode dielectric barrier discharge plasma in air promoting apoptotic behavior in melanoma skin cancer cell lines. *Plasma Chem. Plasma Process.* **2007**, *27*, 163. [CrossRef]
48. Volotskova, O.; Hawley, T.S.; Stepp, M.A.; Keidar, M. Targeting the cancer cell cycle by cold atmospheric plasma. *Sci Rep. UK* **2012**, *2*. [CrossRef] [PubMed]
49. Kim, C.-H.; Bahn, J.H.; Lee, S-H.; Kim, G.-Y.; Jun, S.-I.; Lee, K.; Baek, S.J. Induction of cell growth arrest by atmospheric non-thermal plasma in colorectal cancer cells. *J. Biotechnol.* **2010**, *150*, 530. [CrossRef] [PubMed]
50. Tanaka, H.; Mizuno, M.; Ishikawa, K.; Nakamura, K.; Kajiyama, H.; Kano, H.; Kikkawa, F.; Hori, M. Plasma activated medium selectively kills glioblastoma brain tumor cells by down-regulating a survival signaling molecule, AKT kinase. *Plasma Med.* **2013**, *1*, 265. [CrossRef]
51. Barekzi, N.; Laroussi, M. Dose-dependent killing of leukemia cells by low-temperature plasma. *J. Phys. D Appl. Phys.* **2012**, *45*, 422002. [CrossRef]
52. Huang, J.; Li, H.; Chen, W.; Lu, G.-H.; Wang, X.-Q.; Zhang, G.-P.; Ostrikov, K.; Wang, P.-Y.; Yang, S.-Z. Dielectric barrier discharge plasma in Ar/O2 promoting apoptosis behavior in A549 cancer cells. *Appl. Phys. Lett.* **2011**, *99*, 253701. [CrossRef]
53. Kim, Y.; Ballato, J.; Foy, P.; Hawkins, T.; Wei, Y.; Li, J.; Kim, S.O. Apoptosis of lung carcinoma cells induced by a flexible optical fiber-based cold microplasma. *Biosens. Bioelectron.* **2011**, *28*, 333. [CrossRef] [PubMed]
54. Tanaka, H.; Mizuno, M.; Ishikawa, K.; Nakamura, K.; Utsumi, F.; Kajiyama, H.; Kano, H.; Maruyama, S.; Kikkawa, F.; Hori, M. Cell survival and proliferation signaling pathways are downregulated by plasma activated medium in glioblastoma brain tumor cells. *Plasma Med.* **2012**, *2*, 207. [CrossRef]
55. Tanaka, H.; Mizuno, M.; Kikkawa, F.; Hori, M. Interactions between a plasma-activated medium and cancer cells. *Plasma Med.* **2016**, *6*, 101. [CrossRef]
56. Mohades, S.; Laroussi, M.; Sears, J.; Barekzi, N.; Razavi, H. Evaluation of the Effects of a Plasma Activated Medium on Cancer Cells. *Phys. Plasmas* **2015**, *22*, 122001. [CrossRef]
57. Mohades, S.; Barekzi, N.; Razavi, H.; Maramuthu, V.; Laroussi, M. Temporal Evaluation of Antitumor Efficiency of Plasma Activated Media. *Plasma Process. Polym.* **2016**, *13*, 1206. [CrossRef]
58. Mohades, S.; Laroussi, M.; Maruthamuthu, V. Moderate Plasma Activated Media Supresses Proliferation and Migration of MDCK Epithelial Cells. *J. Phys. D Appl. Phys.* **2017**, *50*, 185205. [CrossRef]
59. Graves, D. The emerging role of reactive oxygen and nitrogen species in redox biology and some implications for plasma applications to medicine and biology. *J. Phys. D Appl. Phys.* **2012**, *45*, 263001. [CrossRef]
60. Lu, X.; Naidis, G.V.; Laroussi, M.; Reuter, S.; Graves, D.B.; Ostrikov, K. Reactive Species in Non-equilibrium Atmospheric Pressure Plasma: Generation, Transport, and Biological Effects. *Phys. Rep.* **2016**, *630*, 1. [CrossRef]
61. Laroussi, M.; Lu, X.; Keidar, M. Perspective: The Physics, Diagnostics, and Applications of Atmospheric Pressure Low Temperature Plasma Sources Used in Plasma Medicine. *J. Appl. Phys.* **2017**, *122*, 020901. [CrossRef]
62. Zhao, S.; Xiong, Z.; Mao, X.; Meng, D.; Lei, Q.; Li, Y.; Deng, P.; Chen, M.; Tu, M.; Lu, X.; et al. Atmospheric Pressure Room Temperature Plasma Jets Facilitate Oxidative and Nitrative Stress and Lead to Endoplasmic Reticulum Stress Dependent Apoptosis in HepG2 Cells. *PLOS ONE* **2013**, *8*, e73665. [CrossRef] [PubMed]
63. Yan, X.; Zou, F.; Zhao, S.; Lu, X.; He, G.; Xiong, Z.; Xiong, Q.; Zhao, Q.; Deng, P.; Huang, J.; et al. On the Mechanism of Plasma Inducing Cell Apoptosis. *IEEE Trans. Plasma Sci.* **2010**, *38*, 2451. [CrossRef]
64. Yan, X.; Xiong, Z.; Zou, F.; Zhao, S.; Lu, X.; Yang, G.; He, G.; Ostrikov, K. Plasma-Induced Death of HepG2 Cancer Cells: Intracellular Effects of Reactive Species. *Plasma Process. Polym.* **2012**, *9*, 59. [CrossRef]
65. Ishaq, M.; Evans, M.; Ostrikov, K. Effects of Atmospheric gas Plasmas on Cancer Cell Signaling. *Int. J. Cancer* **2014**, *134*, 1517. [CrossRef] [PubMed]

66. Ishaq, M.; Kumar, S.; Varinli, H.; Han, Z.J.; Rider, A.E.; Evans, M.; Murphy, A.B.; Ostrokov, K. Atmospheric Gas Plasma-Induced ROS Production Activates TNS-ASK1 Pathway for the Induction of Melanoma Cancer Cell Apoptosis. *Mol. Biol. Cells* **2014**, *25*, 1523. [CrossRef] [PubMed]

67. Oh, J.-S.; Szili, E.J.; Gaur, N.; Hong, S.-H.; Futura, H.; Kurita, H.; Mizuno, A.; Hatta, A.; Short, R.D. How to assess the plasma delivery of RONS into tissue fluid and tissue. *J. Phys. D Appl. Phys.* **2016**, *49*, 304005. [CrossRef]

68. Hong, S.-H.; Szili, E.J.; Toby, A.; Jenkins, A.; Short, R.D. Ionized gas (plasma) delivery of reactive oxygen species (ROS) into artificial cells. *J. Phys. D Appl. Phys.* **2014**, *47*, 362001. [CrossRef]

69. Szili, E.J.; Bradley, J.W.; Short, R.D. A 'tissue model' to study the plasma delivery of reactive oxygen species. *J. Phys. D Appl. Phys.* **2014**, *47*, 152002. [CrossRef]

70. Gaur, N.; Szili, E.J.; Oh, J.; Hong, S.; Michelmore, A.; Graves, D.B.; Hatta, A.; Short, R.D. Combined effect of protein and oxygen on reactive oxygen and nitrogen species in the plasma treatment of tissue. *Appl. Phys. Lett.* **2015**, *107*, 103703. [CrossRef]

71. He, T.; Liu, D.; Xu, H.; Liu, Z.; Xu, D.; Li, D.; Li, Q.; Rong, M.; Kong, M. A 'tissue model' to study the barrier effects of living tissues on the reactive species generated by surface air discharge. *J. Phys. D Appl. Phys.* **2016**, *49*, 205204. [CrossRef]

72. Duan, J.; Lu, X.; He, G. On the penetration depth of reactive oxygen and nitrogen species generated by a plasma jet through real biological tissue. *Phys. Plasmas* **2017**, *24*, 073506. [CrossRef]

73. Begum, A.; Laroussi, M.; Pervez, M.R. Atmospheric Pressure helium/air plasma Jet: Breakdown Processes and Propagation Phenomenon. *AIP Adv.* **2013**, *3*, 062117. [CrossRef]

74. Sobota, A.; Guaitella, O.; Garcia-Caurel, E. Experimentally obtained values of electric field of an atmospheric pressure plasma jet impinging on a dielectric surface. *J. Phys. D Appl. Phys.* **2013**, *46*, 372001. [CrossRef]

75. Stretenovic, G.B.; Krstic, I.B.; Kovacevic, V.V.; Obradovic, A.M.; Kuraica, M.M. Spatio-temporally resolved electric field measurements in helium plasma jet. *J. Phys. D Appl. Phys.* **2014**, *47*, 102001. [CrossRef]

76. Lu, Y.; Wu, S.; Cheng, W.; Lu, X. Electric field measurements in an atmospheric-pressure microplasma jet using Stark polarization emission spectroscopy of helium atom. *Eur. Phys. J. Spec. Top.* **2017**, *226*, 2979. [CrossRef]

77. Pervez, M.R.; Begum, A.; Laroussi, M. Plasma Based Sterilization: Overview and the Stepwise Inactivation Process of Microbial by Non-thermal Atmospheric Pressure Plasma Jet. *Int. J. Eng. Technol.* **2014**, *14*, 7.

78. Kogelschatz, U. Dielectric-Barrier Discharges: Their History, Discharge Physics, and Industrial Applications. *Plasma Chem. Plasma Proc.* **2003**, *23*, 1. [CrossRef]

79. Brandenburg, R. Dielectric barrier discharges: progress on plasma sources and on the understanding of regimes and single filaments. *Plasma Sources Sci. Technol.* **2017**, *26*, 053001. [CrossRef]

80. Laroussi, M.; Lu, X. Room Temperature Atmospheric Pressure Plasma Plume for Biomedical Applications. *Appl. Phys. Lett.* **2005**, *87*, 113902. [CrossRef]

81. Lu, X.; Laroussi, M.; Puech, V. On Atmospheric Pressure Non-equilibrium Plasma Jets and Plasma Bullets. *Plasma Sources Sci. Technol.* **2012**, *21*, 034005. [CrossRef]

82. Weltmann, K.-D.; Kindel, E.; von Woedtke, T.; Hähnel, M.; Stieber, M.; Brandenburg, R. Atmospheric-pressure plasma sources: Prospective tools for plasma medicine. *Pure Appl. Chem.* **2010**, *82*, 1223. [CrossRef]

83. Teschke, M.; Kedzierski, J.; Finantu-Dinu, E.G.; Korzec, D.; Engemann, J. High Speed Photographs of a Dielectric Barrier Atmospheric Pressure Plasma Jet. *IEEE Trans. Plasma Sci.* **2005**, *33*, 310. [CrossRef]

84. Lu, X.; Laroussi, M. Dynamics of an Atmospheric Pressure Plasma Plume Generated by Submicrosecond Voltage Pulses. *J. Appl. Phys.* **2006**, *100*, 063302. [CrossRef]

85. Mericam-Bourdet, N.; Laroussi, M.; Begum, A.; Karakas, E. Experimental Investigations of Plasma Bullets. *J. Phys. D Appl. Phys.* **2009**, *42*, 055207. [CrossRef]

86. Sands, B.L.; Ganguly, B.N.; Tachibana, K. A Streamer-like Atmospheric Pressure Plasma Jet. *Appl. Phys. Lett.* **2008**, *92*, 151503. [CrossRef]

87. Karakas, E.; Koklu, M.; Laroussi, M. Correlation between helium mole fraction and plasma bullet propagation in low temperature plasma jets. *J. Phys. D Appl. Phys.* **2010**, *43*, 155202. [CrossRef]

88. Boeuf, J.-P.; Yang, L.; Pitchford, L. Dynamics of guided streamer (plasma bullet) in a helium jet in air at atmospheric pressure. *J. Phys. D Appl. Phys.* **2013**, *46*, 015201. [CrossRef]

89. Naidis, G. Modeling of streamer propagation in atmospheric-pressure helium plasma jets. *J. Phys. D Appl. Phys.* **2010**, *43*, 402001. [CrossRef]

90. Yousfi, M.; Eichwald, O.; Merbahi, N.; Jomma, N. Analysis of ionization wave dynamics in low-temperature plasma jets from fluid modeling supported by experimental investigations. *Plasma Sources Sci. Technol.* **2012**, *21*, 045003. [CrossRef]

91. Breden, D.; Miki, K.; Raja, L.L. Self-consistent 2D modeling of cold atmospheric-pressure plasma jets/bullets. *Plasma Sources Sci. Technol.* **2012**, *21*, 034011. [CrossRef]

92. Lu, X.; Naidis, G.; Laroussi, M.; Ostrikov, K. Guided Ionization Waves: Theory and Experiments. *Phys. Rep.* **2014**, *540*, 123. [CrossRef]

93. Laroussi, M. Low temperature plasma jet for biomedical applications: A review. *IEEE Trans. Plasma Sci.* **2015**, *43*, 703. [CrossRef]

94. Szili, E.J.; Gaur, N.; Hong, S.-H.; Kurita, H.; Oh, J.-S.; Ito, M.; Mizuno, A.; Hatta, A.; Cowin, A.J.; Graves, D.B.; et al. The assessment of cold atmospheric plasma treatment of DNA in synthetic models of tissue fluid, tissue and cells. *J. Phys. D Appl. Phys.* **2017**, *50*, 274001. [CrossRef]

95. Ray, A.; Ranieri, P.; Karamchand, L.; Yee, B.; Foster, J.; Kopleman, R. Real-Time Monitoring of Intracellular Chemical Changes in Response to Plasma Irradiation. *Plasma Med.* **2017**, *7*, 7. [CrossRef]

 *plasma*

Article

Treatment of Triple-Negative Breast Cancer Cells with the Canady Cold Plasma Conversion System: Preliminary Results

Xiaoqian Cheng [1,†], Warren Rowe [1,†], Lawan Ly [1], Alexey Shashurin [2], Taisen Zhuang [3], Shruti Wigh [3], Giacomo Basadonna [1,4], Barry Trink [1,5], Michael Keidar [1,5] and Jerome Canady [1,5,*]

[1] Jerome Canady Research Institute for Advanced Biological and Technological Sciences, Takoma Park, MD 20912 USA; xcheng@usmedinnov.com (X.C.); drwrowe@usmedinnov.com (W.R.); llawan@usmedinnov.com (L.L.); giacomo.basadonna@umassmed.edu (G.B.); barrytrink@gmail.com (B.T.); keidar@gwu.edu (M.K.)
[2] School of Aeronautics and Astronautics, Purdue University, West Lafayette, IN 47907, USA; ashashur@purdue.edu
[3] Plasma Medicine Life Sciences, Takoma Park, MD 20912, USA; tzhuang@usmedinnov.com (T.Z.); swigh@usmedinnov.com (S.W.)
[4] Department of Surgery, University of Massachusetts School of Medicine, Worcester, MA 01655, USA
[5] School of Engineering and Applied Science, The George Washington University, Washington, DC 20052, USA
* Correspondence: drjcanady@usmedinnov.com; Tel.: +1-301-270-0147
† The authors have contributed equally.

Received: 27 July 2018; Accepted: 10 September 2018; Published: 15 September 2018

Abstract: Triple-negative breast cancer is a phenotype of breast cancer where the expression level of estrogen, progesterone and human epidermal growth factor receptor 2 (HER2) receptors are low or absent. It is more frequently diagnosed in younger and premenopausal women, among which African and Hispanic have a higher rate. Cold atmospheric plasma has revealed its promising ant-cancer capacity over the past two decades. In this study, we report the first cold plasma jet delivered by the Canady Cold Plasma Conversion Unit and characterization of its electric and thermal parameters. The unit effectively reduced the viability of triple-negative breast cancer up to 80% without thermal damage, providing a starting point for future clinical trials.

Keywords: triple-negative breast cancer; cold atmospheric plasma; cold plasma device

1. Introduction

Breast cancer is the most common cancer diagnosed among US women (excluding skin cancers) and is the second leading cause of cancer death among women after lung cancer [1]. Triple-negative breast cancer refers to the breast cancer phenotype which has an absence or low level expression of estrogen, progesterone, and HER2 receptors [2]. It is known for its poor clinical outcome and lack of effective targeted therapy because women with triple-negative breast cancer do not benefit from endocrine therapy or trastuzumab. Chemotherapy is currently the mainstay of systemic medical treatment [3]. Patients with triple-negative disease have a lower three-year survival rate following chemotherapy than patients with breast cancers of other subtypes [4].

Cold atmospheric plasma (CAP) has been extensively studied for its biomedical use in various fields such as surface decontamination [5], wound healing [6,7], dental treatment [8], allergen destruction [9], HIV virus treatment [10] and among others [11]. In particular, the research of CAP as a potential oncotherapeutic approach has thrived over the past decade and the mechanism is been increasingly understood [12–16]. It is widely reported that CAP deactivated more than 20 types of cancer in vitro by inducing apoptosis [17–19], cell cycle arrest [20–22], endoplasmic reticulum stress [23,24] and DNA damage [25–27]. CAP has also been shown to significantly reduce tumor volume in an in vivo murine model following s phase cell cycle arrest and apoptosis [28].

With such a promising anti-cancer capacity, there is as yet to be a commercialized clinical applicable cold atmospheric plasma device reported. This study presents a new invention of an integrated cold plasma and high-frequency plasma electrosurgical system (U.S. Patent No. 9.999,462) [29] allowing tumor removal and treatment of surgical margins in a single device, and evaluates the response of triple-negative breast cancer to treatment with the device. Our study demonstrates CAP, delivered by the Canady Cold Plasma Conversion System, as a potential adjuvant for triple-negative breast cancer treatment.

2. Materials and Methods

2.1. Cell Culture, Treatment, and Viability Assay

All experiments were performed at the Jerome Canady Research Institute for Advanced Biological and Technological Sciences, in Takoma Park, MD, USA. Human breast cancer cell line MDA-MB-231 was generously donated by Professor Michael Keidar's group at The George Washington University. Cells were cultured in Dulbecco's Modified Eagle Medium (DMEM) supplemented with 10% fetal bovine serum and 1% Pen Strep in a 37 °C and 5% CO_2 humidified incubator (Thermo Fisher Scientific, Waltham, MA, USA). When cells reached approximately 80% confluence, cells were seeded at a concentration of 10^5 cells/well into 12-well plates (USA Scientific, Ocala, FL, USA) or 5×10^3 cells/well into 96-well plates (USA Scientific, Ocala, FL, USA). Helium flow was set to a constant 1 lpm at power set 20 P or 40 P on the USMI SS-601 MCa (USMI, Takoma Park, MD, USA) or 3 lpm and power set to 40 P, 60 P, or 80 P. The plasma scalpel was placed 1.5 cm (at 1 lpm) or 2 cm (at 3 lpm) away from the surface of the cell media. Well-plates were placed on a plate heater (Benchmark, New York, NY, USA) which maintained temperature at 37 °C, providing a relatively warmer and gentle environment for the cells during treatment. Thiazolyl Blue Tetrazolium Bromide (MTT) assay was performed on the cells 48 h after plasma treatment following the manufacturer's protocol. All the MTT assay reagents were purchased from Sigma-Aldrich (St. Louis, MO, USA). The absorbance of the dissolved compound was measured by BioTek Synergy HTX (Winooski, VT, USA) microplate reader at 570 nm.

2.2. Cold Plasma Device Power and Temperature Measurement

Electric parameters of the cold plasma discharge were measured using Tektronix P6021A (Tektronix, Beaverton, OR, USA) current probe with a frequency range of 120 Hz to 60 MHz and PPE 6 kV high voltage probe (LeCroy, Chestnut Ridge, NY, USA) attached to a digital oscilloscope Wavesurfer 3024 (LeCroy, Chestnut Ridge, NY, USA). Helium flow rates at 1 lpm and 3 lpm were measured at different power settings. A thermal camera (Wilsonville, OR, USA) was used to collect temperature data. The volume of the media in each well was 1 mL for 12-well plate and 0.1 mL for 96-well plate. The distance between the tip of the scalpel and the surface of the media was kept at 1.5 cm (at 1 lpm) or 2 cm (at 3 lpm). Temperature measurement of the CAP-treated media was also performed with the plate heater set to 37 °C, which was consistent with cell viability experiments. The temperature of the beam and treated media was measured every minute from 0 min (immediately after the CAP was on) to 5 min.

2.3. Statistics

All viability assays were repeated for at least three times with two replicates each time. Data was plotted by Microsoft Excel 2016 (Redmond, WA, USA) as mean ± standard error of the mean. Student *t*-test or one-way analysis of variance (ANOVA) were used to check statistical significance where applicable. Differences were considered statistically significant for * $p \leq 0.05$.

3. Results

3.1. The Canady Cold Plasma Conversion System

The system designed for cancer treatment in this study is reported in detail in a US patent [29]. Briefly, it is comprised of two units, namely the conversion unit (CU, USMI, Takoma Park, MD, USA) and the cold plasma probe (CPP, USMI, Takoma Park, MD, USA). The CU is integrated with a USMI SS-601 MCa, a high-frequency electrosurgical generator (ESU, USMI, Takoma Park, MD, USA) unit and converts the ESU signal. The CPP is connected to the CU output. Plasma is produced at the end of the CPP and is thermally harmless to living tissue, i.e., it is cold plasma. The connection schematics are shown in Figure 1A. The CU is equipped with three connectors, namely a gas connector (to a helium tank), an electrical connector (to ESU), and an electro-gas connector (to CPP). The CU utilizes a high voltage transformer connected to the output from the ESU. The CU up-converts voltage up to 4 kV, down-converts frequency to less than 300 kHz, and down-converts power to less than 40 W. The CPP is connected to an electro-gas output connector of the CU and has a length of 0.5 m. Figure 1B shows the assembly of the whole CAP generation system.

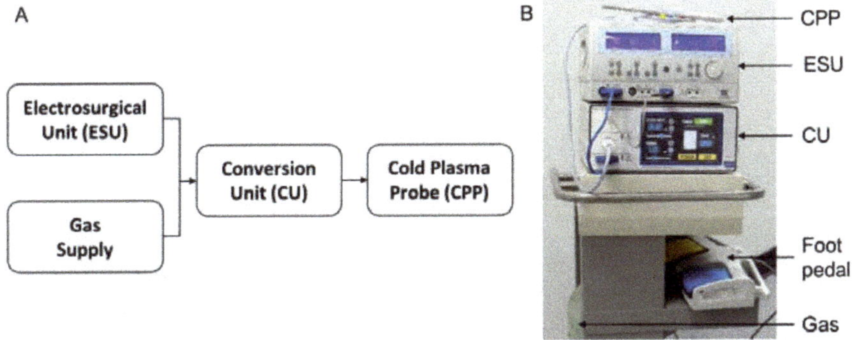

Figure 1. (**A**) Schematics of a system for producing cold plasma by converting a high-frequency electrosurgical unit; (**B**) Picture of the CAP generation system.

3.2. Power and Temperature Measurement of CAP

The electric parameters of the cold plasma discharge were measured and a schematic image of the setup is shown in Figure 2A. Figure 2B shows the output voltage of the ESU (orange line) and the CU (blue line). The ESU spray mode is a pulse modulated system. The ESU generated series of high voltage bursts with peak amplitude of about 1 kV separated about 30 μs between the bursts. Each voltage burst was filled with harmonic oscillations at a frequency of about 880 kHz. The CU output waveform had a smaller resonate frequency about 140 kHz and amplitude about 1–1.5 kV. That is to say, the CU is not a power generating device, but a frequency and voltage modulation device.

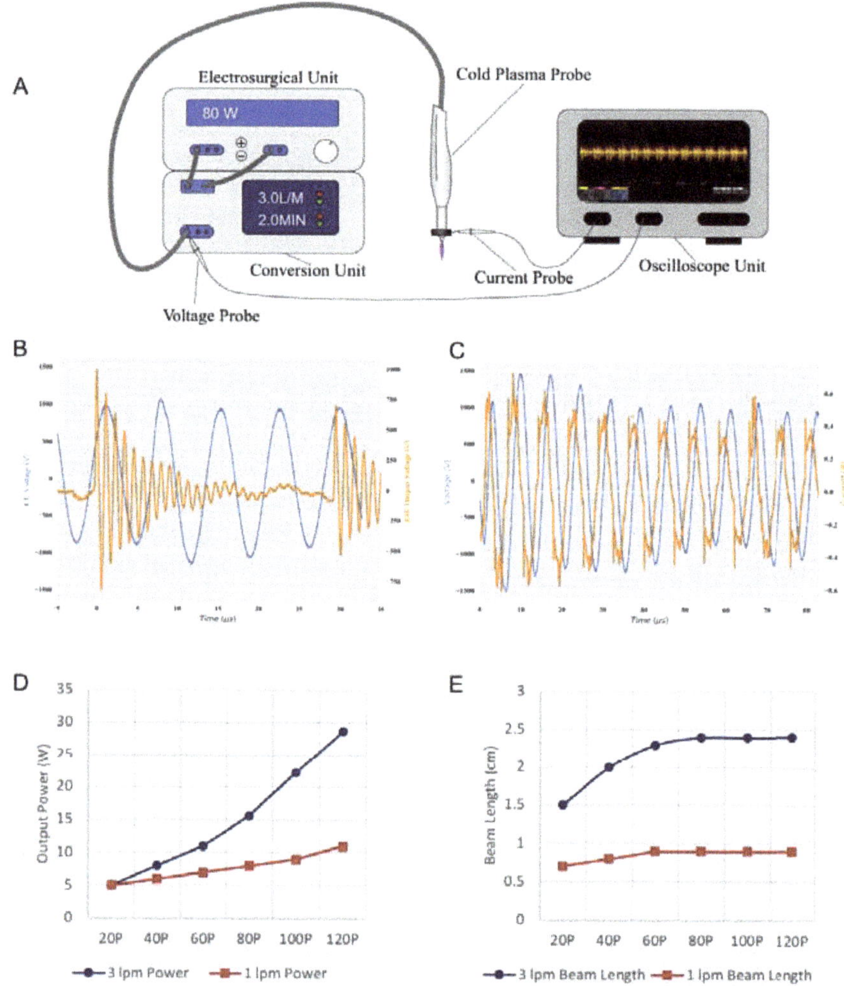

Figure 2. Output power was measured across the plasma scalpel tip. (**A**) Schematic image of power testing setup; (**B**) Output voltage of the ESU (orange line) and the CU (blue line); (**C**) Output voltage and current signals from the CU; (**D**) Power measurement of the CU; (**E**) Beam length of the CAP jet.

The output voltage and current signals from the CU using 3 lpm are shown in Figure 2C. The blue line indicates the voltage output from the CU and the orange line is the current dissipate through the CPP, with the ESU set to 60 P. We observed a phase shift between voltage and current curve; the current curve is about 80 degrees ahead of the voltage curve. In other words, the cold plasma system is acting as neither a pure resistive nor a pure reactive impedance. Therefore, the real power delivered to the discharge was calculated as follows. The power deposited into the cold plasma discharge was calculated by the oscilloscope directly as $\frac{1}{T}\int_T U \cdot I\, dt$ for large integration time $T = 2$ ms (over 20 M data points, more than 200 oscillations). The power deposited into the cold plasma discharge at 20 P, 40 P, 60 P, 80 P, 100 P, and 120 P for 3 lpm and 1 lpm was plotted in Figure 2D. The power settings of 20 P, 40 P, 60 P, 80 P, 100 P, and 120 P yield powers deposited into the cold plasma discharge of 5 W, 8 W, 11 W, 15.7 W, 22.3 W, and 28.7 W at 3 lpm respectively; 5 W, 6 W, 7 W, 8 W, 9W, and 11 W at 1 lpm respectively.

The length of the CAP beam was also measured at different power settings for both flow rates and plotted in Figure 2E. At 3 lpm, the length of the beam increases rapidly from 1.5 to 2 cm and to 2.3 cm when the power is increased from 20 P to 40 P and to 60 P. After 60 P, the length remains constant at 2.4 cm as the power further increases up to 120 P. This trend also applies to 1 lpm. The length of the beam increases from 0.7 to 0.8 cm when the power increases from 20 P to 40 P, and maintains a length of 0.9 cm for 80 P, 100 P, and 120 P.

Thermal images of the CAP jet with flow rates of 3 lpm and 1 lpm are shown in Figures 3A and 4A respectively. Cell culture media was warmed up to 37 °C beforehand and added to well plates immediately before measurement. The environment temperature was about 23 °C during the experiment. As shown in the graphs, for both flow rates the temperature of the treated media (Figures 3B and 4B) as well as the CAP beam (Figures 3C and 4C) increases with power increasing from 20 P to 120 P. The beam temperature of 3 lpm CAP is about 26 to 30 °C, whereas the treated area of the 12-well plate is 15 to 21 °C. In the case of 1 lpm, the beam temperature is in the range of 25.5 to 31 °C, and the treatment area is roughly 23 to 36 °C.

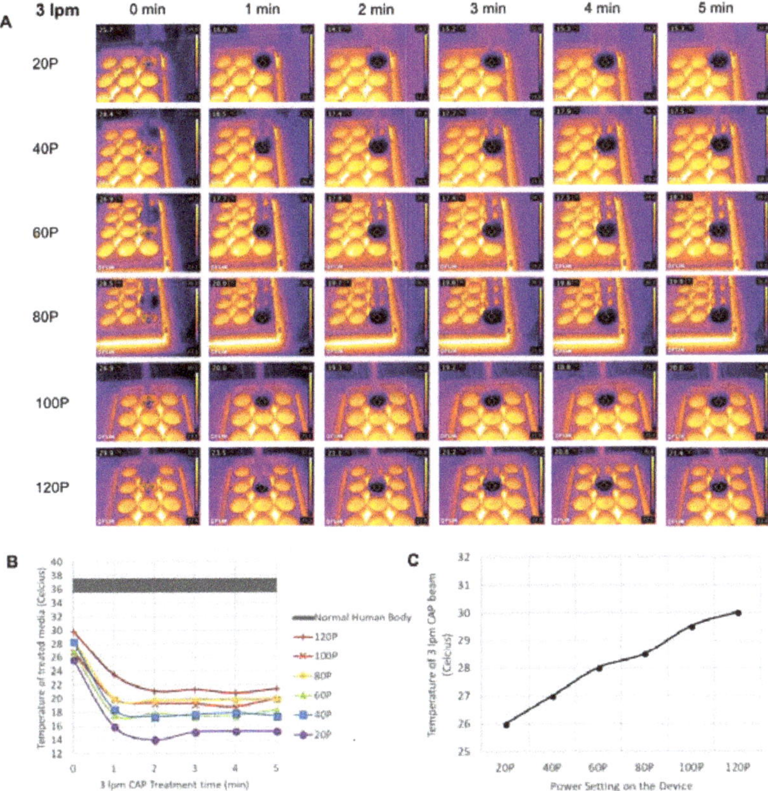

Figure 3. Temperature measurement at flow rate of 3 lpm for each power setting (**A**) Thermal images of CAP-treated media; (**B**) Temperature of CAP-treated media as a function of treatment time; (**C**) Temperature of the CAP beam.

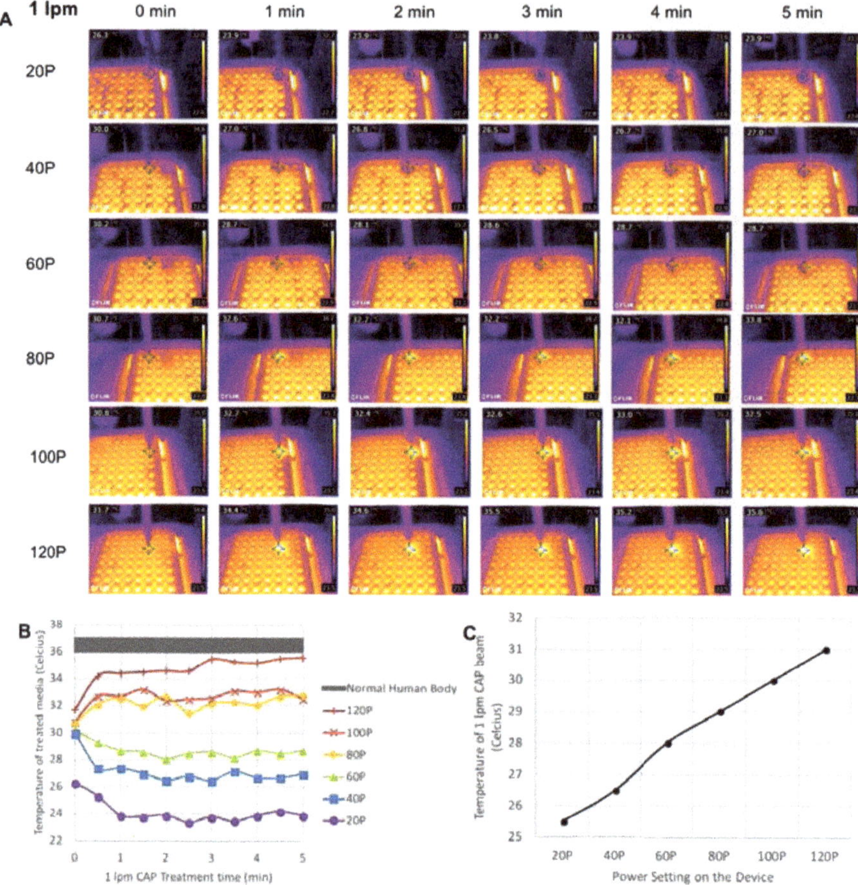

Figure 4. Temperature measurement at flow rate of 1 lpm for each power setting (**A**) Thermal images of CAP-treated media; (**B**) Temperature of CAP-treated media as a function of treatment time; (**C**) Temperature of the CAP beam.

3.3. Cell Viability after CAP Treatment

Cells were treated by the Canady Cold Plasma Conversion System and viability was quantified by MTT assay 48 h after CAP treatment. As shown in Figure 5A,B primary axis, CAP treatment significantly reduced the proliferation of triple-negative cancer cells at various conditions. CAP treatment of MDA-MB-231 significantly reduces viability at nearly all doses tested using 3 lpm (Figure 5A). At 1 lpm flow rate, 90 to 120 s of CAP treatment was needed to significantly reduce viability (Figure 5B).

The energy delivered by the CU to the CPP can be calculated as

$$E = P \times t \tag{1}$$

where E is the total energy of the CAP (J) delivered by the system; P is the power measured at the end of the CPP (W), and t is the treatment time (s). The consumed energy of each CAP treatment condition used in this study was plotted as the secondary axis of Figure 5A (3 lpm) and 5B (1 lpm). The reduction of cell viability matches the energy consumption for both flow rates and this trend is consistent across all power and time settings tested. A picture of the cells treated by the CPP is shown in Figure 5C.

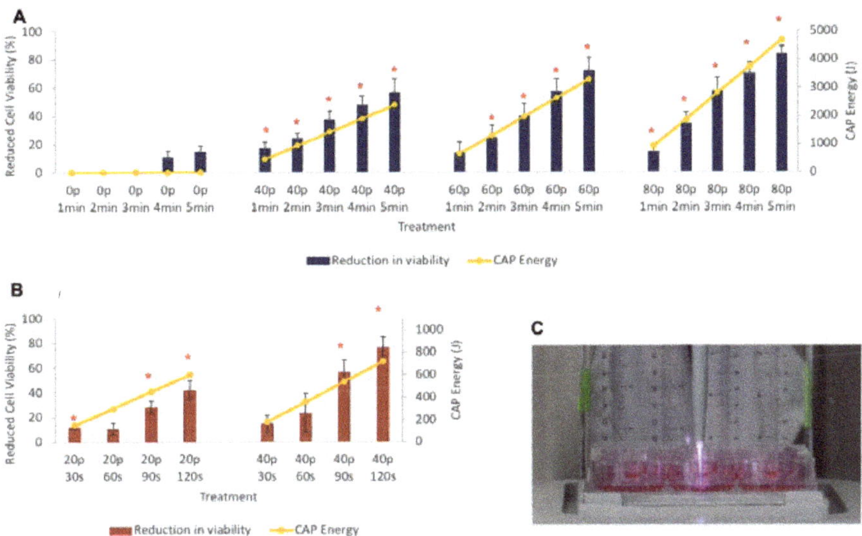

Figure 5. Reduced viability of treatment on MDA-MB-231 measured by MTT assay (bar chart, primary axis) and energy deposited in the corresponding CAP treatment (line chart, secondary axis) (**A**) Cells were treated by CAP at 3 lpm in 12-well plates; (**B**) Cells were treated by CAP at 1 lpm in 96-well plates; (**C**) Picture of 3 lpm CAP jet treating the cells in a 12-well plate. * P $\leq$ 0.05.

4. Discussion

Cold plasma can be generated in various forms including dielectric barrier discharge, corona discharge, and plasma jets [30]. The Canady Cold Plasma Conversion Unit reported in this study is the first cold plasma device that utilizes a high voltage transformer to up-convert the voltage, down-convert the frequency, and down-convert the power of the high voltage output from an electrosurgical unit (U.S. Patent No. 9.999,462) [29].

The plasma jet generated by the Canady Cold Plasma Conversion System is indeed "cold". The beam temperature for all conditions tested is within the range of 26 to 31 °C. It has a cooling effect on the treated media when the flow rate is high and/or power is low. With a flow rate at 3 lpm, the beam temperature is 26 to 30 °C for 20 P to 120 P, whereas the treated media in the 12-well plate is about 15 to 21 °C respectively.

The beam temperature of 1 lpm CAP is 25.5 to 31 °C for 20 P to 120 P, which is very close to that of 3 lpm. Although the power parameters of CAP are higher at 3 lpm than 1 lpm, as shown above in the Results Section, the similar temperature could be resulted from better heat convection of the higher flow. For a lower power setting of 20 P to 60 P, the temperature of the CAP-treated media in the 96-well plate at 1 lpm, 24 to 29 °C, is lower or close to the CAP beam temperature. However, for the higher power setting of 80 P to 120 P, the temperature of the treated media in the 96-well plate, 32 to 36 °C, is 3 to 5 degrees higher than the beam temperature. Theoretically, the media temperature should only increase to the beam temperature based on the principle of heat transfer. However, during the experiments we observed that the CAP jet was intensified, which could be caused by the energy dissipating to the wall of the 96-well plate due to the high power as well as the turbulence resulting from a small well size. The increased intensity of the CAP jet is demonstrated as high brightness in the thermal images in Figure 4A Row 4 to Row 6 (80 P to 120 P). The length of the CAP beam also presents evidence of increased intensity. As shown above in Figure 2C, the beam is only 0.9 cm at 80 P to 120 P for 1 lpm when measured in open air, while the thermal images in Figure 4A were captured with 1.5 cm gap distance between the CPP tip and the media. This disparity is because when treating in the well, even at lower power settings, the beam is able to reach 1.5 cm and contact the media. The 3 lpm CAP jet does not present this issue because the diameter of a 12-well plate is significantly larger than the CAP jet.

When applied to cells, power settings of 20 P to 80 P for 3 lpm and 20 P to 40 P for 1 lpm were chosen to ensure the integrity of the CAP delivered to the cells. The temperature of the treatment area is between 15 to 30 °C for all treatment conditions at all times, suggesting no thermal damage to the cells.

The CAP generated by the Canady Cold Plasma Conversion System affects triple-negative breast cancer in a power- and time-dependent manner which corresponds with the increased output power and beam length shown in Figure 2C. The CAP reduced the viability of triple-negative breast cancer up to 80% at the highest power for both flow rates. To further illustrate the correspondence between cell viability and CAP power and treatment time, we calculate the energy delivered by the system; E. The close correlation between energy consumption and reduction in viability may be important for comparing results between different CAP devices. Difference cell types may also respond differently to CAP treatment, therefore more studies have been performed to confirm the efficacy of the CAP system on other solid tumor cancer cell lines [31] as well as the safety of the unit on normal tissue [32]. Future animal studies are needed to determine the optimal dosage for cancer elimination while remaining safe for normal tissue.

Although at 3 lpm, the CAP jet delivers higher energy than 1 lpm with the same power and time setting, the MTT assay shows a similar reduction in viability (Figure 5A,B). Thus, direct comparison of the cell viability cannot be made between 3 lpm and 1 lpm despite the same power setting and treatment time due to different beam length, well size, medium volume, and cell number between the two conditions. This was also reported repeatedly by other researchers worldwide [33,34]. This study intends to focus on the introduction of the Canady Cold Plasma Conversion System and its efficacy on the treatment of triple-negative breast cancer for both flowrates.

To better understand the strength of each cold plasma dosage and evaluate the efficiency of the cold plasma beam during treatment, cell viability reduction rate (CVRR) was introduced in this paper. CVRR was calculated based on the cell viability rate versus the time at a constant ESU power setting (%/s). Figure 6 shows the CVRR values and averaged CVRR corresponding to 40 P, 60 P and 80 P ESU power setting at 3 lpm. There is little difference when comparing the CVRR between treatment times and the average only increases slightly with increased power. In other words, the average CVRR value could represent the overall performance of the CAP for that power setting. More importantly, one could establish a treatment projected solution based on the average CVRR value, and in that sense, CVRR is an ideal parameter to calculate the CAP dose.

The poor prognosis and low overall survival rate of triple-negative breast cancer demands a novel and safe treatment. The high-frequency converted cold plasma system integrates coagulation and CAP in a single device, making it more practical for medical applications. After the surgeon removes the cancerous tumor, CAP is subsequently sprayed at the surgical margins to ablate any remaining cancerous tissue or cells, thus reducing the chances of cancer recurrence. CAP treatment acts as an important adjunct to the current treatment protocol for solid cancerous tumors. This new plasma system will change the landscape of electrosurgery and cancer therapy as well as offer cancer patients new hope in the very near-future.

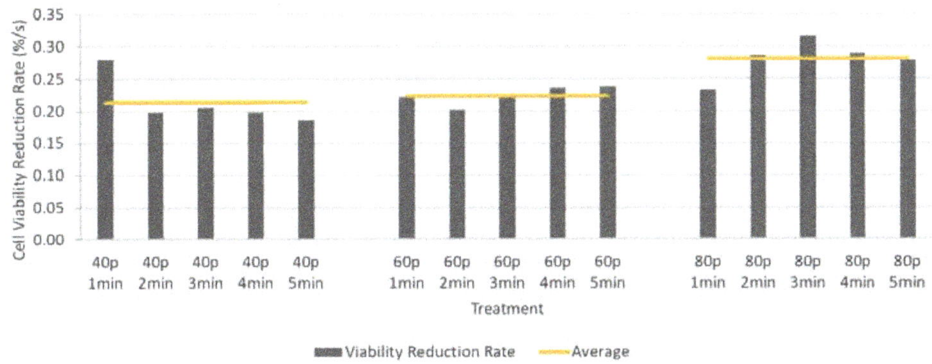

Figure 6. Reduction rate of MDA-MB-231 cells by CAP treatment under different conditions.

5. Conclusions

This study reports the first cold plasma jet delivered by the Canady Cold Plasma Conversion Unit, which was characterized and tested on triple-negative breast cancer cells. Viability of these cells was effectively reduced in a time- and power-dependent manner. Our CAP device, a direct conversion from a high-frequency electrosurgical unit, allows for the removal of a tumor and the following treatment by CAP for ablating cancer cells using a single device, and this study will contribute to the dosage estimation for patients in future clinical applications.

Author Contributions: Conceptualization, X.C., W.R., G.B., B.T., M.K. and J.C.; Data curation, X.C., W.R., L.L. and T.Z.; Formal analysis, X.C., W.R. and T.Z.; Funding acquisition, J.C.; Investigation, X.C., W.R., L.L., T.Z. and S.W.; Methodology, X.C., W.R., A.S. and M.K.; Project administration, X.C. and W.R.; Software, X.C., W.R. and T.Z.; Supervision, M.K. and J.C.; Validation, X.C. and W.R.; Writing—original draft, X.C., W.R. and T.Z.; Writing—review & editing, X.C., W.R., A.S., T.Z., G.B., B.T., M.K. and J.C.

Funding: This research received no external funding.

Acknowledgments: The authors would like to thank the engineer team at Plasma Medicine Life Sciences for technical support of the plasma unit. This research was funded by US Medical Innovations, LLC.

Conflicts of Interest: The authors declare no conflict of interest.

References

1. DeSantis, C.E.; Ma, J.; Goding Sauer, A.; Newman, L.A.; Jemal, A. Breast cancer statistics, 2017, racial disparity in mortality by state. *CA-Cancer J. Clin.* **2017**, *67*, 439–448. [CrossRef] [PubMed]

2. Wahba, H.A.; El-Hadaad, H.A. Current approaches in treatment of triple-negative breast cancer. *Cancer Biol. Med.* **2015**, *12*, 106–116. [PubMed]

3. Foulkes, W.D.; Smith, I.E.; Reis-Filho, J.S. Triple-negative breast cancer. *New Engl. J. Med.* **2010**, *363*, 1938–1948. [CrossRef] [PubMed]

4. Liedtke, C.; Mazouni, C.; Hess, K.R.; Andre, F.; Tordai, A.; Mejia, J.A.; Symmans, W.F.; Gonzalez-Angulo, A.M.; Hennessy, B.; Green, M.; et al. Response to neoadjuvant therapy and long-term survival in patients with triple-negative breast cancer. *J. Clin. Oncol.* **2008**, *26*, 1275–1281. [CrossRef] [PubMed]

5. Niemira, B.A.; Boyd, G.; Sites, J. Cold plasma rapid decontamination of food contact surfaces contaminated with salmonella biofilms. *J. Food Sci.* **2014**, *79*, M917–M922. [CrossRef] [PubMed]

6. Schmidt, A.; Woedtke, T.V.; Stenzel, J.; Lindner, T.; Polei, S.; Vollmar, B.; Bekeschus, S. One year follow-up risk assessment in skh-1 mice and wounds treated with an argon plasma jet. *Int. J. Mol. Sci.* **2017**, *18*, 868. [CrossRef] [PubMed]

7. Schmidt, A.; Wende, K.; Bekeschus, S.; Bundscherer, L.; Barton, A.; Ottmuller, K.; Weltmann, K.D.; Masur, K. Non-thermal plasma treatment is associated with changes in transcriptome of human epithelial skin cells. *Free Radical Res.* **2013**, *47*, 577–592. [CrossRef] [PubMed]

8. Pierdzioch, P.; Hartwig, S.; Herbst, S.R.; Raguse, J.D.; Dommisch, H.; Abu-Sirhan, S.; Wirtz, H.C.; Hertel, M.; Paris, S.; Preissner, S. Cold plasma: A novel approach to treat infected dentin-a combined ex vivo and in vitro study. *Clin. Oral Investig.* **2016**, *20*, 2429–2435. [CrossRef] [PubMed]

9. Wu, Y.; Liang, Y.; Wei, K.; Li, W.; Yao, M.; Zhang, J. Rapid allergen inactivation using atmospheric pressure cold plasma. *Environ. Sci. Technol.* **2014**, *48*, 2901–2909. [CrossRef] [PubMed]

10. Volotskova, O.; Dubrovsky, L.; Keidar, M.; Bukrinsky, M. Cold atmospheric plasma inhibits hiv-1 replication in macrophages by targeting both the virus and the cells. *PLoS ONE* **2016**. [CrossRef] [PubMed]

11. Bußler, S.; Herppich, W.B.; Neugart, S.; Schreiner, M.; Ehlbeck, J.; Rohn, S.; Schlüter, O. Impact of cold atmospheric pressure plasma on physiology and flavonol glycoside profile of peas (pisum sativum "Salamanca"). *Food Res. Int.* **2015**, *76*, 132–141. [CrossRef]

12. Yan, D.; Sherman, J.H.; Keidar, M. Cold atmospheric plasma, a novel promising anti-cancer treatment modality. *Oncotarget* **2017**, *8*, 15977–15995. [CrossRef] [PubMed]

13. Keidar, M. Plasma for cancer treatment. *Plasma Sources Sci. Technol.* **2015**, *24*, 20. [CrossRef]

14. Laroussi, M.; Lu, X.; Keidar, M. Perspective: The physics, diagnostics, and applications of atmospheric pressure low temperature plasma sources used in plasma medicine. *J. Appl. Phys.* **2017**. [CrossRef]

15. Schlegel, J.; Köritzer, J.; Boxhammer, V. Plasma in cancer treatment. *Clin. Plasma Med.* **2013**, *1*, 2–7. [CrossRef]

16. Tanaka, H.; Ishikawa, K.; Mizuno, M.; Toyokuni, S.; Kajiyama, H.; Kikkawa, F.; Metelmann, H.; Hori, M. State of the art in medical applications using non-thermal atmospheric pressure plasma. *Rev. Mod. Plasma Phys.* **2017**, *1*, 89. [CrossRef]

17. Ishaq, M.; Han, Z.J.; Kumar, S.; Evans, M.D.M.; Ostrikov, K.K. Atmospheric-pressure plasma- and trail-induced apoptosis in trail-resistant colorectal cancer cells. *Plasma Processes Polym.* **2015**, *12*, 574–582. [CrossRef]

18. Adachi, T.; Tanaka, H.; Nonomura, S.; Hara, H.; Kondo, S.; Hori, M. Plasma-activated medium induces a549 cell injury via a spiral apoptotic cascade involving the mitochondrial-nuclear network. *Free Radical Biol. Med.* **2015**, *79*, 28–44. [CrossRef] [PubMed]

19. Weiss, M.; Gumbel, D.; Hanschmann, E.M.; Mandelkow, R.; Gelbrich, N.; Zimmermann, U.; Walther, R.; Ekkernkamp, A.; Sckell, A.; Kramer, A.; et al. Cold atmospheric plasma treatment induces anti-proliferative effects in prostate cancer cells by redox and apoptotic signaling pathways. *PLoS ONE* **2015**. [CrossRef] [PubMed]

20. Shi, X.; Zhang, G.; Chang, Z.; Wu, X.; Liao, W.; Li, N. Viability reduction of melanoma cells by plasma jet via inducing G1/S and G2/M cell cycle arrest and cell apoptosis. *IEEE Trans. Plasma Sci.* **2014**, *42*, 1640–1647. [CrossRef]

21. Gherardi, M.; Turrini, E.; Laurita, R.; De Gianni, E.; Ferruzzi, L.; Liguori, A.; Stancampiano, A.; Colombo, V.; Fimognari, C. Atmospheric non-equilibrium plasma promotes cell death and cell-cycle arrest in a lymphoma cell line. *Plasma Processes Polym.* **2015**, *12*, 1354–1363. [CrossRef]

22. Volotskova, O.; Hawley, T.S.; Stepp, M.A.; Keidar, M. Targeting the cancer cell cycle by cold atmospheric plasma. *Sci. Rep.* **2012**. [CrossRef] [PubMed]

23. Ruwan Kumara, M.H.; Piao, M.J.; Kang, K.A.; Ryu, Y.S.; Park, J.E.; Shilnikova, K.; Jo, J.O.; Mok, Y.S.; Shin, J.H.; Park, Y.; et al. Non-thermal gas plasma-induced endoplasmic reticulum stress mediates apoptosis in human colon cancer cells. *Oncol. Rep.* **2016**, *36*, 2268–2274. [CrossRef] [PubMed]

24. Zhao, S.; Xiong, Z.; Mao, X.; Meng, D.; Lei, Q.; Li, Y.; Deng, P.; Chen, M.; Tu, M.; Lu, X.; et al. Atmospheric pressure room temperature plasma jets facilitate oxidative and nitrative stress and lead to endoplasmic reticulum stress dependent apoptosis in HepG2 cells. *PLoS ONE* **2013**. [CrossRef] [PubMed]

25. Zhang, X.; Zhang, C.; Zhou, Q.Q.; Zhang, X.F.; Wang, L.Y.; Chang, H.B.; Li, H.P.; Oda, Y.; Xing, X.H. Quantitative evaluation of DNA damage and mutation rate by atmospheric and room-temperature plasma (ARTP) and conventional mutagenesis. *Appl. Microbiol. Biotechnol.* **2015**, *99*, 5639–5646. [CrossRef] [PubMed]

26. Chung, W.H. Mechanisms of a novel anticancer therapeutic strategy involving atmospheric pressure plasma-mediated apoptosis and DNA strand break formation. *Arch. Pharmacal Res.* **2016**, *39*, 1–9. [CrossRef] [PubMed]

27. Chang, J.W.; Kang, S.U.; Shin, Y.S.; Kim, K.I.; Seo, S.J.; Yang, S.S.; Lee, J.S.; Moon, E.; Baek, S.J.; Lee, K.; et al. Non-thermal atmospheric pressure plasma induces apoptosis in oral cavity squamous cell carcinoma: Involvement of DNA-damage-triggering sub-G1 arrest via the ATM/p53 pathway. *Arch. Biochem. Biophys.* **2014**, *545*, 133–140. [CrossRef] [PubMed]

28. Vandamme, M.; Robert, E.; Lerondel, S.; Sarron, V.; Ries, D.; Dozias, S.; Sobilo, J.; Gosset, D.; Kieda, C.; Legrain, B.; et al. Ros implication in a new antitumor strategy based on non-thermal plasma. *Int. J. Cancer* **2012**, *130*, 2185–2194. [CrossRef] [PubMed]

29. Canady, J.; Shashurin, A.; Keidar, M.; Zhuang, T. Integrated cold plasma and high frequency plasma electrosurgical system and method. U.S. Patent 9,999,462, 19 June 2018.

30. Weltmann, K.D.; Kindel, E.; von Woedtke, T.; Hähnel, M.; Stieber, M.; Brandenburg, R. Atmospheric-pressure plasma sources: Prospective tools for plasma medicine. *Pure Appl. Chem.* **2010**, *82*, 1223–1237. [CrossRef]

31. Rowe, W.; Cheng, X.; Ly, L.; Zhuang, T.; Basadonna, G.; Trink, B.; Keidar, M.; Canady, J. The canady helios cold plasma scalpel significantly decreases viability in malignant solid tumor cells in a dose-dependent manner. *Plasma* **2018**, *1*, 177–188. [CrossRef]

32. Ly, L.; Jones, S.; Shashurin, A.; Zhuang, T.; Rowe, W.; Cheng, X.; Wigh, S.; Naab, T.; Keidar, M.; Canady, J. A new cold plasma jet: Performance evaluation of cold plasma, hybrid plasma and argon plasma coagulation. *Plasma* **2018**, *1*, 189–200. [CrossRef]

33. Yan, D.; Talbot, A.; Nourmohammadi, N.; Cheng, X.; Canady, J.; Sherman, J.; Keidar, M. Principles of using cold atmospheric plasma stimulated media for cancer treatment. *Sci. Rep.* **2015**. [CrossRef] [PubMed]

34. Xu, X.; Dai, X.; Xiang, L.; Cai, D.; Xiao, S.; Ostrikov, K. Quantitative assessment of cold atmospheric plasma anti-cancer efficacy in triple-negative breast cancers. *Plasma Processes Polym.* **2018**. [CrossRef]

Article

Plasma Treatment of Ovarian Cancer Cells Mitigates Their Immuno-Modulatory Products Active on THP-1 Monocytes

Sander Bekeschus [1,*], Can Pascal Wulf [1,2], Eric Freund [1,2], Dominique Koensgen [2], Alexander Mustea [2], Klaus-Dieter Weltmann [1] and Matthias B. Stope [3]

[1] ZIK plasmatis, Leibniz Institute for Plasma Science and Technology, 17489 Greifswald, Germany; can.wulf@yahoo.de (C.P.W.); eric.freund@inp-greifswald.de (E.F.); weltmann@inp-greifswald.de (K.-D.W.)

[2] Department of Gynaecology and Obstetrics, University Medical Center Greifswald, 17475 Greifswald, Germany; koensgen@uni-greifswald.de (D.K.); alexander.mustea@uni-greifswald.de (A.M.)

[3] Department of Urology, University Medical Center Greifswald, 17475 Greifswald, Germany; stopem@uni-greifswald.de

* Correspondence: sander.bekeschus@inp-greifswald.de; Tel.: +49-3834-554-3948

Received: 1 August 2018; Accepted: 14 September 2018; Published: 15 September 2018

Abstract: Cancers modulate their microenvironment to favor their growth. In particular, monocytes and macrophages are targeted by immuno-modulatory molecules installed by adjacent tumor cells such as ovarian carcinomas. Cold physical plasma has recently gained attention as innovative tumor therapy. We confirmed this for the OVCAR-3 and SKOV-3 ovarian cancer cell lines in a caspase 3/7 independent and dependent manner, respectively. To elaborate whether plasma exposure interferes with their immunomodulatory properties, supernatants of control and plasma-treated tumor cells were added to human THP-1 monocyte cultures. In the latter, modest effects on intracellular oxidation or short-term metabolic activity were observed. By contrast, supernatants of plasma-treated cancer cells abrogated significant changes in morphological and phenotypic features of THP-1 cells compared to those cultured with supernatants of non-treated tumor cell counterparts. This included cell motility and morphology, and modulated expression patterns of nine cell surface markers known to be involved in monocyte activation. This was particularly pronounced in SKOV-3 cells. Further analysis of tumor cell supernatants indicated roles of small particles and interleukin 8 and 18, with MCP1 presumably driving activation in monocytes. Altogether, our results suggest plasma treatment to alleviate immunomodulatory secretory products of ovarian cancer cells is important for driving a distinct myeloid cell phenotype.

Keywords: kINPen; plasma medicine; tumor immunology; ovarian cancer

---

1. Introduction

In the field of plasma medicine, cells and tissues are exposed to partially ionized gas plasma for therapeutic effects [1]. Cold physical plasmas operated at body temperature (i.e., cause no thermal damage) and are potentially bio-active through a number of components generated including reactive oxygen and nitrogen species (ROS/RNS), ions and electrons, ultraviolet (UV) radiation, and electrical fields. Over the last two decades, potential applications have extended from eradicating microorganisms during wound healing to inactivating tumor cells [2]. Today, effective killing of the latter has been demonstrated for different types of cancers including head and neck [3–5], leukemia [6–8], glioblastoma [9–11], pancreas [12–14], malignant melanoma [15–17], colon [18–20], prostate [21–23], osteosarcoma [24–26], and ovarian [27–29]. It has been established the ROS/RNS are the main drivers of antitumor plasma effects [30–32].

With 161,100 deaths worldwide in 2015 and a 5-year survival rate of 45%, ovarian cancer ranks 8th in the list of deaths from cancer [33]. The peritoneum of ovarian cancer patients is characterized by an influx of immune cells including monocytes and macrophages [34]. The presence of macrophages also characterizes the bulk tumor [35], presumably due to the early recruitment of monocytes from the blood [36]. As in other tumor types, macrophages are skewed by ovarian cancer cells towards a tumor-supporting phenotype that provides growth factors and fosters angiogenesis [37]. Tumor-associated macrophages (TAMs) are closely associated with hypoxic tumor lesions [38]. ROS can reprogram TAMs to a more inflammatory macrophage phenotype. This illustrates the importance of the tumor cell-monocyte/macrophage axis as potential therapeutic target [39].

Soluble mediators derived from malignant as well as non-malignant cells dictate the inflammatory status within the tumor microenvironment, myeloid differentiation responses, and the level of peritoneal inflammation key in the metastasis of ovarian cancer [40–42]. Using cold physical plasma as potent source of ROS for ovarian cancer cell inactivation, here we investigated the immunomodulatory role of their secretion products in human THP-1 monocytes. Using the two cancer cell lines OVCAR-3 and SKOV-3, we were able to demonstrate that plasma treatment, at least in part, reverted a tumor cell-induced monocyte/macrophage phenotype.

2. Materials and Methods

2.1. Cell Culture, Plasma Treatment, and Supernatants

The human ovarian cancer cell lines OVCAR-3 (ATCC HTB-161) and SKOV-3 (HTB-77) as well as human THP-1 monocytes (ATCC TIB-202) were maintained in Roswell Park Memorial Medium (RPMI) supplemented with 2% glutamine, 1% penicillin/streptomycin, and 10% FCS (all Sigma, Taufkirchen, Germany). For plasma treatment of ovarian cancer cell lines, 2×10^5 cells were suspended in 500 µL of fully supplemented medium and added to 24-well plates (Eppendorf, Hamburg, Germany). Cells were exposed for 30 s to cold physical plasma of the atmospheric pressure plasma jet kINPen (neoplas, Greifswald, Germany) with argon (air liquid, Paris, France) as feed gas at a flux of four standard liters per minute. Control cells were left untreated or exposed to argon gas alone for 30 s. After 4 h of incubation, supernatants of several wells pooled, centrifuged at $1000\times g$ for 5 min to eradicate any remaining cells, and stored at -20 °C until further use. THP-1 cells were not directly exposed to plasma in this study. Instead, they were cultured either in fully supplemented medium or in fully supplemented medium containing supernatants of control or plasma-treated ovarian cancer cells (80 µL of medium or supernatants with 20 µL of THP-1 cell suspension). 1×10^4 THP-1 cells were cultured in 96-well plates (Eppendorf) with a water-filled rim to prevent evaporation during incubation for up to 96 h. An extra of 50 µL cell culture medium was added to THP-1 cells after 48 h for long-term cultures.

2.2. Metabolic Activity and Viability of Ovarian Cancer Cells

To assess metabolic activity of ovarian cancer cells after plasma treatment they were incubated with resazurin (Alfa Aesar, Haverhill, MA, USA) at a final concentration of 100 µM. After 4 h of incubation, supernatants were transferred to a flat-bottom 96-well plate. Fluorescence was assessed using a microplate reader (Tecan, Männedorf, Switzerland) at λ_{ex} 535 nm and λ_{em} 590 nm, and normalized to that of untreated control cells. For apoptosis at 4 h, cells were incubated with caspase 3/7 reagent (life technologies, Darmstadt, Germany) and 4′,6-Diamidin-2-phenylindol (DAPI, Sigma) for 15 min. Subsequently, cells were harvested and caspase and DAPI fluorescence was measured at a single-cell level using flow cytometry (CytoFlex; Beckman-Coulter, Brea, CA, USA). For time resolved cell death analysis, DAPI$^+$ cells were quantified over a time course of 4 h using fluorescence microscopy.

2.3. High Content Imaging of THP-1 Monocytes

Imaging experiments were performed using an Operetta CLS high content imaging device (PerkinElmer, Hamburg, Germany), and analyzed with Harmony 4.6 software (PerkinElmer). To investigate motility of THP-1 cells, time-lapse video microscopy was performed at a 15 min interval for 4 h. Cells were segmented, boarder objects were removed, and tracking kinetics for individual cells were calculated. For calculation of mean cell-based cytosolic area, imaging was performed using digital phase contrast at 24 h. Phorbol-12-myristat-13-acetat (PMA; Sigma) at a final concentration of 100 nM served as positive control. To generate a morphometric analysis of THP-1 cells, segmented objects were analyzed using the software-specific SER HOLE feature that allows texture-based read out of pixel distributions in the bright field channel at 24 h in one experiment. For all types of quantitative image analysis, 9–36 fields of views (FoV) were acquired for each of three to six technical replicates per condition. The sum of segmentation-based object count in these FoV was, depending on the assay type and incubation time, approximately 500–3000 that was used to generate mean values for the respective analysis. Approximately 4000 images were analyzed in this study.

2.4. Metabolic Activity and Flow Cytometry of THP-1 Monocytes

Metabolic activity of THP-1 cells cultured with or without tumor cell supernatants was assessed at 24 h following a 4 h incubation period with resazurin as described above for ovarian cancer cells. Total cell counts were retrieved by flow cytometry (CytoFlex) at 24 h. To detect immediate oxidative effects THP-1 cells, the latter were stained with hydroxyphenyl fluorescein (HPF) and mitotracker orange (MTO), or CM-H_2-DCF-DA (DCF; all life technologies) as well as DAPI prior to exposure to tumor culture supernatants. This specific mitotracker dye only accumulates in cells with intact mitochondrial membrane potential ($\Delta\Psi$m). Thereafter, mean fluorescence intensities in viable (DAPI-negative) cells were assessed by flow cytometry, and normalized to that of THP-1 cells, which had received fully supplemented culture medium only. Hydrogen peroxide (H_2O_2) served as positive control for DCF staining. Cell surface marker analysis was performed at 96. For this, THP-1 cell culture medium was collected and any remaining THP-1 cells attaching to the plastic bottom of the 96-well plate were detached and added to the respective matching tube. Cells were washed with phosphate-buffered saline (PBS; Sigma) and incubated for 15 min in the dark with fluorochrome-labelled antibodies targeted against the following surface markers: CD15s (PerCP Cy5.5), CD33 (BV510), CD41 (Pacific Blue), CD45RA (PE-Dazzle), CD49d (FITC), CD55 (PE-Cy5), CD63 (PE), CD69 (BV650), CD154 (APC7), CD271 (PE-Cy7), and HLA-ABC (APC) (all BioLegend, London, United Kingdom). PMA served as positive control. After two subsequent washes with PBS, cells were analyzed by 11-color flow cytometry. Only viable cells (with appropriate forward scatter and side scatter properties) were included in the analysis. Gating and compensation was performed using Kaluza analysis software 2.1 (Beckman-Coulter). Approximately 500 individual flow cytometry measurements were included in this study.

2.5. Supernatant Analysis

Tumor cell supernatants were stained with Bodipy (life technologies) and analyzed for small particle release by flow cytometry (Gallios, Beckman-Coulter) as previously described [43] in one experiment with several technical replicates. Heat-shock protein 27 (HSP27) was measured using enzyme-linked immunosorbent assay (ELISA, RnD Systems, Wiesbaden, Germany) according to the manufacturer's instructions. Multiplex cytokine analysis of THP-1 (at day 6) and tumor cell supernatants (at 4 h) was performed with the LegendPlex bead-array based quantification kit (BioLegend) according to the vendor's instructions. Beads were analyzed by flow cytometry. 5-log logarithmic functions and LegendPlex software 8.0 (Vigenetech, Carlisle, USA) were employed to calculate absolute target concentrations. Several technical replicates of pooled supernatants from at least three independent experiments were used for ELISA and multiplex cytokine analysis.

2.6. Statistical Analysis

For each assay if not indicated otherwise, at least three independent experiments with at least three technical replicates each were performed and included into data analysis. Unpaired student's T-test was employed to compare the effect of argon or plasma treatment to untreated control or to medium control, as well as between untreated and plasma treatment. Prism 7.05 software (GraphPad software, La Jolla, CA, USA) was employed for data calculation (mean and standard error), graphing, and statistical analysis. Level of significance is indicated as follows: α = 0.05 (*), α = 0.01 (**), α = 0.001 (***).

3. Results

3.1. Plasma Treatment Abrogated Growth of OVCAR-3 and SKOV-3 Cells

Before analyzing the effect of supernatants from control and plasma-treated tumor cell on THP-1 cells, the toxic amplitude of the plasma exposure was tested. Four hours after plasma treatment, the metabolic activity of both OVCAR-3 and SKOV-3 cells was significantly reduced compared to untreated control (Figure 1a). Argon gas treatment gave no effect. SKOV-3 cells were significantly more sensitive to plasma treatment than OVCAR-3. To understand the kinetics on cell demise, properties of both cell types was follow by time-lapse fluorescence microscopy (Figure 1b). Terminal cell death (DAPI$^+$ cells) was increasing with elapsed time and more pronounced in SKOV-3 compared to OVCAR-3 cells. Staining for actives caspases 3 and 7 at 4 h, enzymes eminent in apoptosis induction, we found an increase in apoptotic cells (Figure 1c) with plasma treatment but this was only significant for SKOV-3 but not OVCAR-3 cells (Figure 1d). Nonetheless, OVCAR-3 showed a significant reduction in metabolic activity following plasma treatment. This suggested the caspase assay possibly suffering from extensive background signal as we repeatedly observed high numbers of active caspase 3/7 positive cells in our controls. Having elaborated the impact of plasma in cancer cell death, supernatants were analyzed for their immunomodulatory activity in THP-1 monocytes, next.

3.2. Supernatants of Plasma-Treated Ovarian Cancer Cell Lines Modestly Affected THP-1 Monocytes' Oxidative Balance and Metabolic Activity

As cancer cell supernatants stemmed from plasma-treated cultures, it was important to assess their remaining oxidative potential. For this, we stained THP-1 cells with a redox-sensitive fluorescent dye CM-H$_2$DCF-DA, and tested its responsiveness to hydrogen peroxide (Figure 2a) in a dose-dependent fashion (Figure 2b). Upon culture with supernatants with OVCAR-3 but not SKOV-3 cells, a significant increase was observed as well (Figure 2c), although on a much lower scale (~5%) compared to lowest positive control (15%, Figure 2a). These results were replicated using a second fluorescent redox indicator, HPF. For mitotracker orange, which selectively accumulates in mitochondria with intact membrane potential, a small but consistent and significant decrease was observed across all samples compared to vehicle control (but not supernatants of control and plasma-treated tumor cells). In some respect, this suggests that tumor cell supernatants generally were sensed different to vehicle culture medium control by THP-1 cells. As mitochondria are important sites of ATP generation, it was subsequently investigated whether the supernatants affected THP-1 metabolic activity and cell count as well the ratio of both (i.e., metabolic activity per cell). Apart from cell counts with SKOV-3 supernatants, no significant changes were observed between treatments or compared to vehicle control (Figure 2d). In tendency, the supernatants of plasma-treated tumor cells reduced THP-1 metabolic activity. Altogether, having noted changes in intracellular oxidation and subtle difference in metabolic activity and growth in THP-1 cells exposed to tumor cell supernatants, microscopic analysis was performed analyze the cells further.

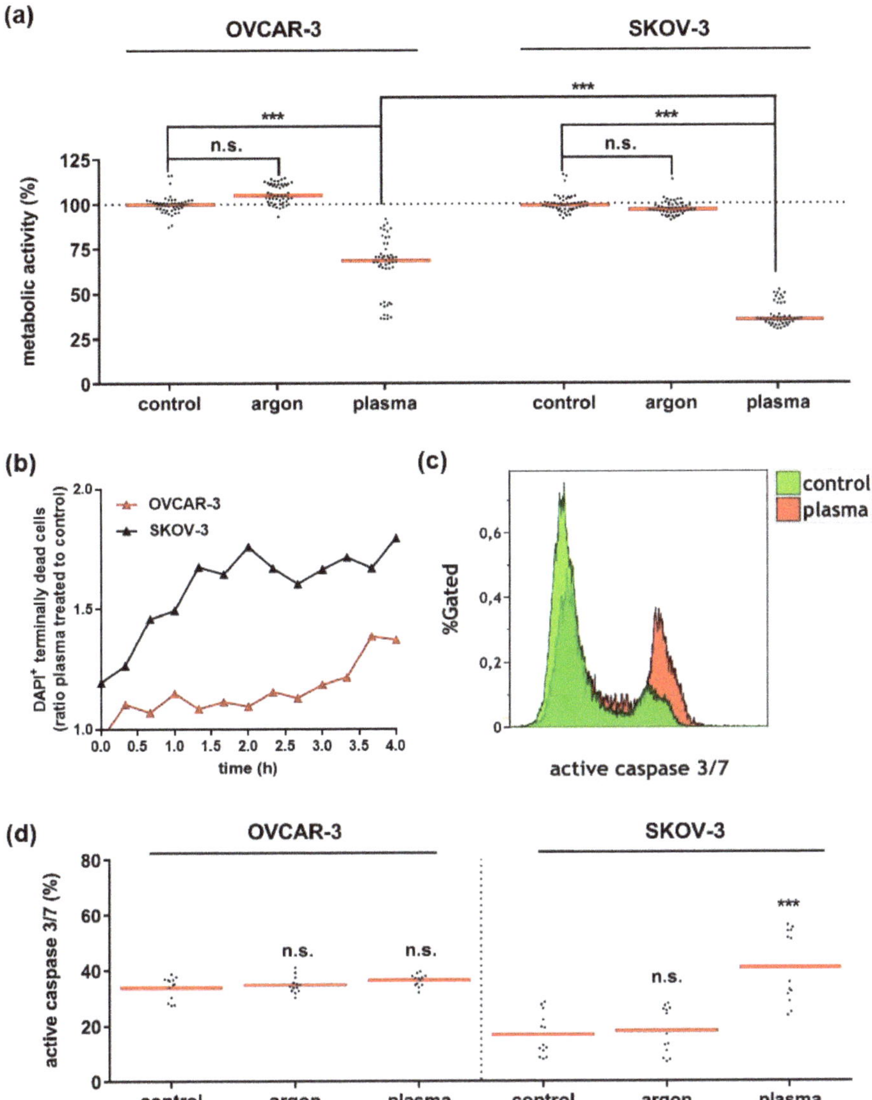

Figure 1. Plasma treatment abrogated growth of OVCAR-3 and SKOV-3 cells. (**a**) Metabolic activity normalized to that of untreated control cells; (**b**) ratio of live over dead cells from quantitative images analysis of 4′,6-Diamidin-2-phenylindol (DAPI⁺) (terminally dead) cells as assayed via time lapse microscopy; (**c**) representative fluorescence histogram overlay of untreated control and plasma-treated SKOV-3 cells for active caspases 3 and 7 assayed via flow cytometry; (**d**) quantification of the percentage of caspase 3/7 positive cells in tumor cells under different conditions. *** ($p < 0.001$) indicates significant difference between indicated samples.

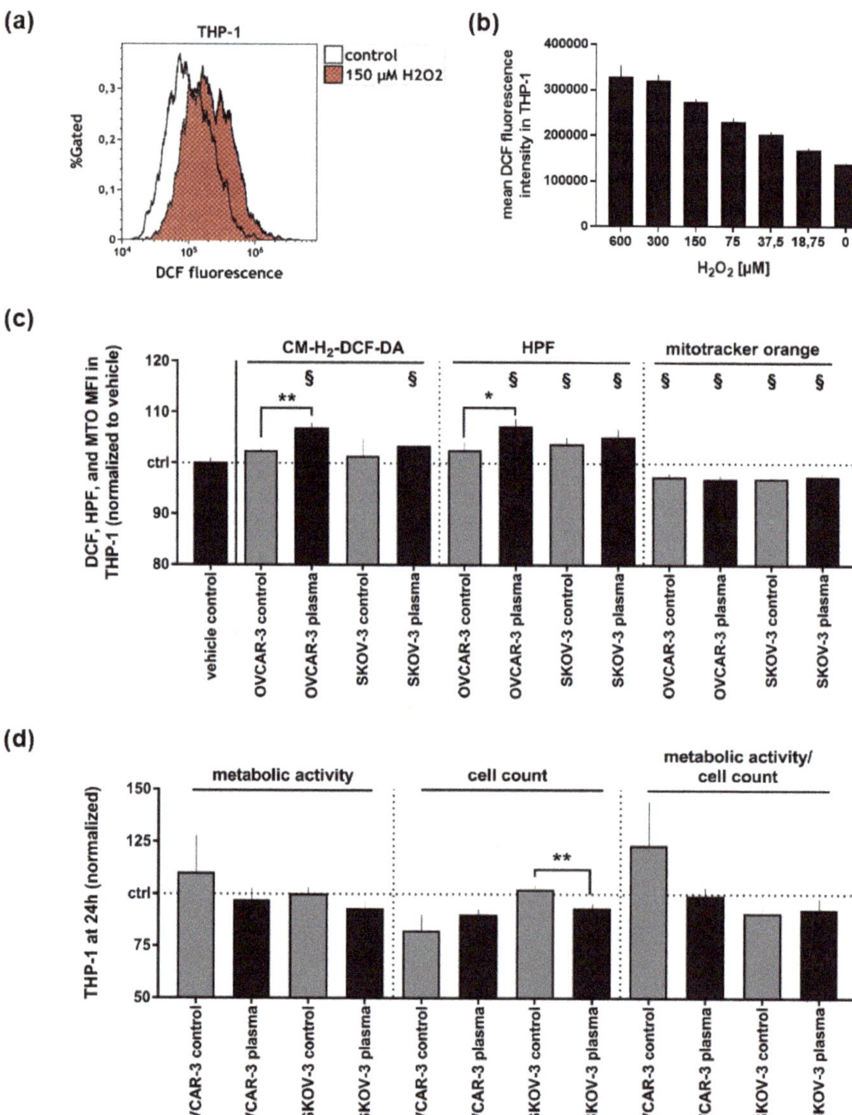

Figure 2. Supernatants of plasma-treated ovarian cancer cell lines modestly affected THP-1 monocytes' oxidative balance and metabolic activity. (**a**) left: representative fluorescence overlay histogram of control and H_2O_2 treated THP-1 cells; (**b**) quantification of mean DCF fluorescence in a H_2O_2 titration series as positive control; (**c**) quantification of DCF, hydroxyphenyl fluorescein (HPF), and mitotracker orange fluorescence intensities of THP-1 cells cultured with tumor cell supernatants or untreated (vehicle) control medium that served as normalization control; (**d**) quantification of metabolic activity, cell count, and metabolic activity over cell count ratio in THP-1 cells cultured with tumor cell supernatants at 24 h, each assay was normalized against untreated control medium. § indicates significant ($p < 0.05$) differences to vehicle control; * ($p < 0.05$) and ** ($p < 0.01$) indicate significant difference between indicated samples.

3.3. Plasma Treatment of Ovarian Cancer Cell Lines Alleviated Effects of Their Supernatants on THP-1 Monocyte Motility and Morphology

Quantitative image analysis is a powerful tool to investigate the morphometric features of cells across a large number of images and individual cells. For tracking the speed of THP-1 monocytes, cells were segmented and tracking tools employed (Figure 3a) to retrieve distance over time data at 4 h. Quantification revealed that supernatants of both OVCAR-3 and SKOV-3 cells decreased cells motility (Figure 3b). Strikingly, plasma-treated supernatants did not do so but rather rescued speed loss back to the level of vehicle control medium. Next, it was assessed whether THP-1 cells co-cultured with tumor cell supernatants displayed morphological features of activated cells. Specifically, non-activated THP-1 cells are round and dense, thus show a high cytosolic intensity when measured with "digital phase contrast" (DPC; Figure 3c, left image). By contrast, activating agents such as PMA lead to a flattening of cells concomitant with a loss of DPC intensity (Figure 3c, right image). Quantification at 24 h revealed that DPC intensity of THP-1 cells cultured with tumor cell supernatants was not different from that of vehicle control, especially in comparison to positive control PMA (Figure 3d). To perform more detailed visual fingerprinting of THP-1 cells, texture based feature analysis (SER HOLE) was employed in bright field images segmented from DPC intensities (Figure 3e). This allows measuring the intracellular density and distribution quantitatively. The white arrow (Figure 3e) shows a denser cell with the correspondingly different SER HOLE image. Analogously to cell motility measurements (Figure 3b), supernatants of plasma-treated tumor cells alleviated the decrease in SER HOLE intensity seen with control tumor-cell supernatants in THP-1 monocytes at 24 h (Figure 3f). In summary, we observed effects of tumor cell supernatants on THP-1 cells motility and morphology, which was reduced to vehicle control levels with plasma-treated supernatants.

3.4. Plasma Treatment of SKOV-3 Cells Mitigated Effects of Their Supernatants on THP-1 Monocyte Surface Marker Expression

Cell surface marker analysis provides a good mean to investigate the activation status of immune cells. We performed a multicolor flow cytometry panel for 11 markers known to be regulated in THP-1 monocyte activation. To validate our approach, we first compared the staining intensity differences between unstained and stained, non-activated cells, which allows assessing baseline marker expression (Figure 4a). Expression of CD41, CD45RA, CD63, and CD69 was weak or absent in resting monocytes. In parallel, we stained PMA-activated monocytes/macrophages and measured marker intensities at 96 h using flow cytometry. Except for CD49d, all markers investigated were upregulated in THP-1 monocytes/macrophages upon PMA treatment (Figure 4a). These measurements helped in judgement of subsequent data from THP-1 monocytes incubated with tumor cell-derived supernatants (Figure 4b). Neither control nor plasma-treated OVCAR-3 supernatants notably modulated the expression of any of the 11 markers in THP-1 monocytes at 96 h (Figure 4c). Conversely, control SKOV-3 supernatants upregulated the expression of eight markers also observed with PMA-induced monocyte-to-macrophage differentiation (Figure 4b), although to a lesser degree. Another marker, CD49d was down regulated, similar to PMA samples. Strikingly, this response was diminished when THP-1 monocytes were incubated with plasma-treated SKOV-3 supernatants. These results suggested that plasma treatment mitigates immunomodulatory effects of tumor cell-derived supernatants in THP-1 monocytes.

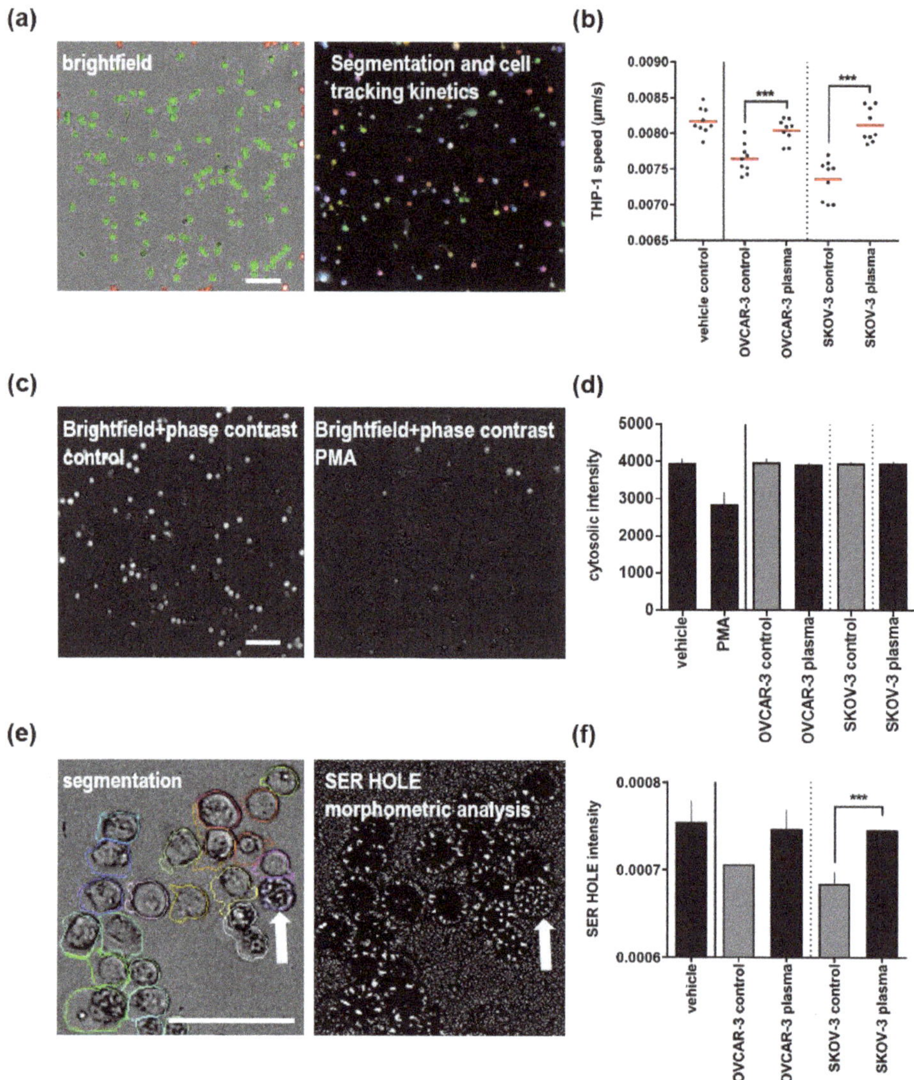

Figure 3. Plasma treatment of ovarian cancer cell lines alleviated effects of their supernatants on THP-1 monocyte motility and morphology. (**a**) Left: representative bright field image with segmented THP-1 cells (green) and excluded boarder objects (red), right: representative digital phase contrast (DPC) image with individual distance accumulation over 4 h; (**b**) quantification of tracking distances in THP-1 cells; (**c**) representative bright field image overlay with DPC of control (left) and PMA-treated (right) THP-1 cells at 24 h, note the decrease in white DPC signal in PMA samples; (**d**) quantification of cytosolic intensity from DPC images; (**e**) segmented THP-1 cells in bright field (left) and SER HOLE feature display (right), note the indicated (white arrow) cell with altered morphometric properties (left) that become a quantitative measure by a less empty (hole) area within the cell (right); (**f**) quantification of SER HOLE intensity. Scale bar = 50 µm.

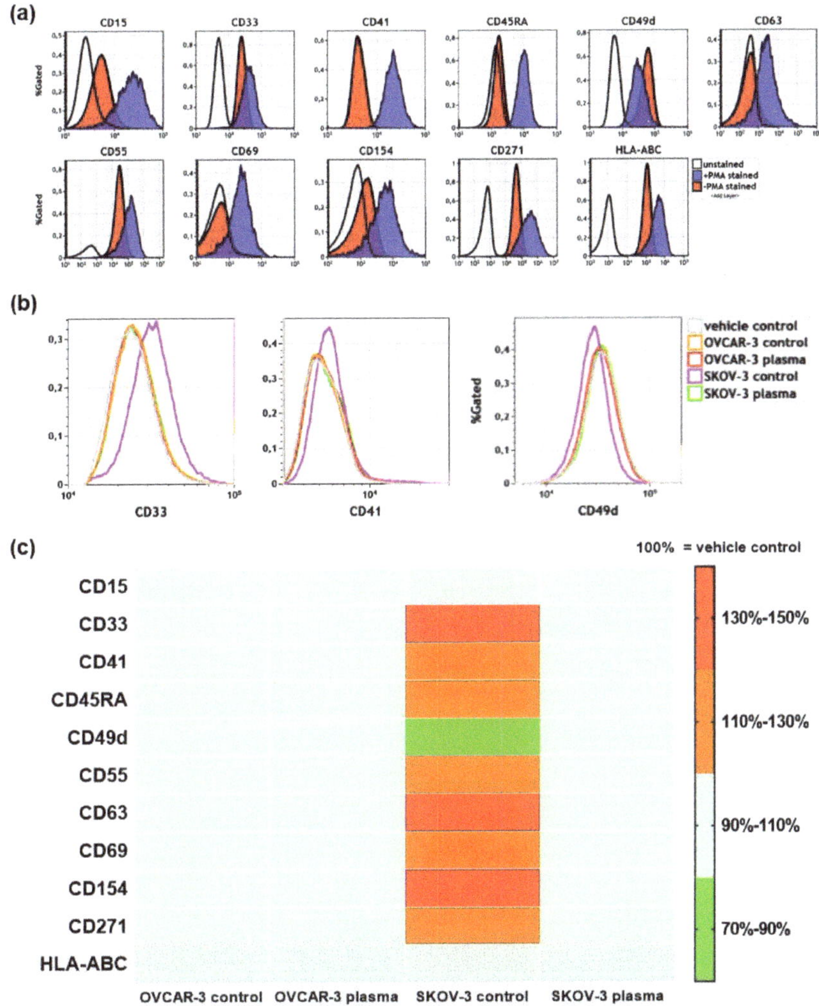

Figure 4. Plasma treatment of SKOV-3 cells mitigated effects of their supernatants on THP-1 monocyte surface marker expression. (**a**) Representative overlay histograms showing the fluorescence intensity of unstained control (white), stained control (red), and stained PMA-treated (blue) THP-1 cells at 96 h; (**b**) representative overlay histograms showing the fluorescence intensities of CD33, CD41, and CD49d in THP-1 cells receiving different types of tumor supernatants; (**c**) summary heat map of fold changes in surface marker expression in THP-1 monocytes with tumor cell supernatants normalized to that of THP-1 cells incubated with vehicle control cell culture medium.

3.5. Plasma Treatment of Ovarian Cancer Cells Changed Their Secretory Products and Those of THP-1 Cells

OVCAR-3 and SKOV-3 supernatants modulated THP-1 activity, morphology, and surface marker expression. To screen for potential mediators, we further investigated these supernatants for potentially active mediators. Many cell types, including cancer cells are capable of modifying inflammation via the release of small vesicles or particles budded from the cell's surface such as microparticles, capable of inducing differentiation-like responses in human monocytes [44]. To this end, we analyzed the cell-depleted supernatants of OVCAR-3 and SKOV-3 cells for membrane-containing vesicles by staining with a bodipy dye and measuring fluorescence intensities via flow cytometry as established previously [43]. Plasma treatment qualitatively (Figure 5a) and quantitatively (Figure 5b) decreased the total number of small particles (range approximately 400 to 3000 μm) measured at 4h in equal liquid volumes. Although the downregulation of small particles may contribute to the effects observed above, we further investigated supernatants for soluble mediators principally released by tumor cells and known to modulate myeloid cell activity, such as heat shock protein 27 (HSP27) [45]. Plasma treatment profoundly increased the amount of HSP27 in supernatants of OVCAR-3 but not SKOV-3 cells (Figure 5c). As these results did not explain the responses seen in THP-1 monocytes, we performed a 13-plex cytokine screening in supernatants of control and plasma-treated tumor cells at 4h as well as THP-1 monocyte supernatants cultured for 96 h in vehicle, control, or plasma-treated tumor cell supernatants (Table 1). In tumor cell supernatants, many analytes were not detected with IL8 being the only strongly expressed chemokine/cytokine found. It was upregulated with plasma treatment, similar to IL18 and TNFα, although to a much lesser extent and only in OVCAR-3 cells. For THP-1 monocytes, only IL8, IL18, and MCP1 were detected. IL8 was absent in vehicle controls but measurable in all samples that have received tumor cell supernatants, possible as endogenous response to the same or as remaining molecules originally stemming from ovarian cancer cells. IL18 was significantly decreased in THP-1 monocytes receiving plasma-treated compared to control OVCAR-3 supernatants, despite the reverse release pattern found in OVCAR-3 supernatants alone. This suggest a THP-1-dependent regulation of IL18 but does not explain the results observed in cell surface marker analysis (Figure 4c). By contrast, MCP1 levels strongly reflected the findings with surface marker expression patterns in monocytes. Levels in supernatants of THP-1 cells receiving vehicle control medium, or control or plasma-treated OVCAR-3 supernatants were similar. Control supernatants of SKOV-3 cells gave a 4-fold increased release in THP-1 cells but a significant decline to vehicle control levels with plasma-treated SKOV-3 supernatants. In summary, reduced amounts of small particles and/or an unknown immunomodulator in ovarian cancer cell supernatants may explain a suggested MCP1-driven activation response in human THP-1 monocytes.

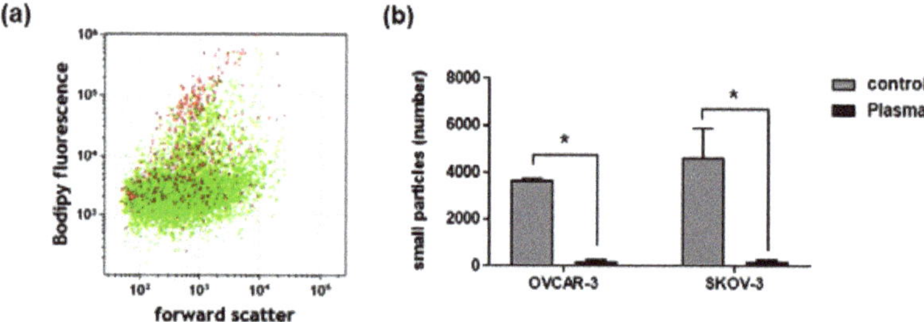

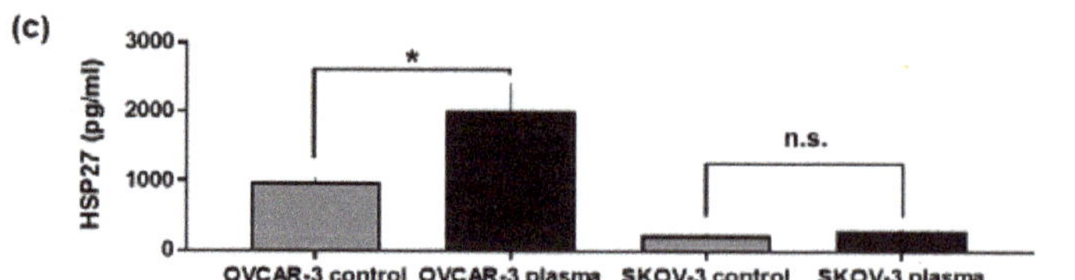

Figure 5. Plasma treatment of ovarian cancer cells changed their secretory products. (**a**) Representative bivariate dot plot of noise-discriminated small particles in the supernatants of control (green) and plasma-treated (red) OVCAR-3 cells at 4 h; (**b**) absolute quantification of small particles present in 5 μL of supernatant; (**c**) quantification of heat-shock-protein 27 (HSP27).

Table 1. 13-plex cytokine analysis of cell culture supernatants retrieved in this study.

Sample Source		THP-1 Monocyte Supernatants					OVCAR-3 Supernatants		SKOV-3 Supernatants	
Target	Statistic	Vehicle	Ov-Ctrl	Ov-Pl	Sk-Ctrl	Sk-Pl	Ctrl	Pl	Ctrl	Pl
IL1β	Mean	2.5	2.5	2.5	2.5	2.5	2.5	2.5	2.5	2.5
	S.D.	0.0	0.0	0.0	0.0	0.0	0.0	0.0	0.0	0.0
IL8	Mean	2.1	**40.2**	**2.7 *****	18.6	3.6	133.3	280.1	**33.8**	**69.1 ****
	S.D.	0.0	0.9	0.8	28.6	2.6	59.4	99.1	6.8	9.6
IL10	Mean	1.2	1.2	1.2	1.2	1.2	1.2	1.2	1.2	1.2
	S.D.	0.0	0.0	0.0	0.0	0.0	0.0	0.0	0.0	0.0
IL12p70	Mean	1.3	1.3	1.3	1.3	1.3	1.3	1.3	1.3	1.3
	S.D.	0.0	0.0	0.0	0.0	0.0	0.0	0.0	0.0	0.0
IL17	Mean	4.2	4.2	4.2	4.2	4.2	4.2	4.2	4.2	4.2
	S.D.	0.0	0.0	0.0	0.0	0.0	0.0	0.0	0.0	0.0
IL18	Mean	3.5	**7.9**	**2.7 ****	2.0	2.2	**7.1**	**29.5 ***	1.3	1.3
	S.D.	1.4	0.4	0.9	1.2	0.8	3.0	8.4	0.0	0.0
IL23	Mean	3.4	3.4	3.4	3.4	3.4	3.4	3.4	3.4	3.4
	S.D.	0.0	0.0	0.0	0.0	0.0	0.0	0.0	0.0	0.0
IL33	Mean	3.0	3.0	3.0	3.0	3.0	3.0	3.0	3.0	3.0
	S.D.	0.0	0.0	0.0	0.0	0.0	0.0	0.0	0.0	0.0
IFNα	Mean	0.7	0.7	0.7	0.7	0.7	0.7	0.7	0.7	0.7
	S.D.	0.0	0.0	0.0	0.0	0.0	0.0	0.0	0.0	0.0
IFNγ	Mean	2.7	2.7	2.7	2.7	2.7	2.7	2.7	2.7	2.7
	S.D.	0.0	0.0	0.0	0.0	0.0	0.0	0.0	0.0	0.0
MCP1	Mean	12.2	14.1	17.3	44.5	**13.6 *****	3.1	3.1	3.1	3.1
	S.D.	2.7	3.2	3.1	2.3	2.3	0.0	0.0	0.0	0.0
TNFα	Mean	1.7	1.7	1.7	1.7	1.7	**1.7**	**4.7 ****	1.7	1.7

IL: interleukin; IFN: interferon; MCP1: monocyte-chemoattractant protein 1 (CCL2); S.D.: standard deviation; Ov: OVCAR-3; Sk: SKOV-3; Ctrl: control; Pl: plasma; *($p < 0.05$), ** ($p < 0.01$), and *** ($p < 0.001$) mark significant differences to the respective control of the matching supernatant and/or cell type.

4. Discussion

Monocyte education and macrophage polarization by ovarian cancer cells is key in tumor progression. Cold physical plasma has been suggested for cancer treatment, and we investigated the impact of secretory products of ovarian cancer cells in human THP-1 monocytes in the context of plasma.

Several lines of evidence argued for THP-1 immunomodulation or possibly activation by ovarian cancer cell supernatants. A decrease in mitochondrial membrane potential as well as an increase in ROS was observed, which is associated with inflammasome-mediated THP-1 cell activation [46]. Moreover, a change in motility and morphological features was observed, events known from earlier studies with THP-1 activation [47]. This was concomitant with a change in surface marker expression patterns at 96 h, although only in one of four supernatant conditions investigated. Possibly, early activation events by tumor cell supernatants were abrogated during long culture conditions for the remaining three. However, we could not identify morphological features at 96 h to confirm findings by flow cytometry as high cell densities at 96 h hampered quantitative image analysis. Cytokine patterns (increase in IL8, IL18, and MCP1 compared to vehicle control) reflected an immunomodulation in THP-1 cells cultured in tumor cell supernatants as well. IL18 is marker of inflammasome activation in THP-1 cells [48], while lipopolysaccharide-stimulated THP-1 monocytes secrete IL8 [49] and MCP1 [50]. IL8 is also released in response to oxidative stimuli [51]. The absence of IL1β and TNFα is in line with findings in primary monocyte-derived macrophages [52]. Although the amplitude of total chemokine/cytokine secretion was rather low, there was a striking similarity between MCP1 release and flow cytometry surface marker profiles. MCP1 is a potent chemoattractant and activator for monocytes [53]. As this chemokine was not present in tumor cell supernatants, we speculate that MCP1 activated THP-1 monocytes in a paracrine or autocrine fashion in response to SKOV-3-derived secretory products.

There are several possibilities for immunomodulatory agents in tumor cell supernatants that, alone or in concert with plasma treatment, could have acted on THP-1 monocytes. A subtle, yet consistent increase in intracellular THP-1 cell oxidation was seen with tumor cell supernatants and elevated in plasma conditions. For the latter, it is tempting to speculate that this finding is owed to remaining plasma-derived reactive species [54] in the supernatants. However, the kINPen generates approximately 60 μM of hydrogen peroxide (H_2O_2) within 30 s of plasma treatment of 500 μL medium [55,56]. With 200,000 cells suspended in the medium, it is unlikely that significant amounts of H_2O_2 would not have reacted with tumor cells across 4 h of incubation 37 °C and after freeze-thawing of the supernatants. Instead, it is possible that remaining ROS were secondary derivatives of tumor cells, possibly as active response to stress [57] or as ROS leaking from damaged mitochondria present during apoptosis [58]. Alternatively, ROS increase in THP-1 may have been due to plasma-induced oxidation in biomolecules [59] taken up the cells or immediate activity increase in response to, for instance, chemokines and cytokines. ROS are vital signaling agents observed during activation of monocytes [60], and the subtle but consistent increase in ROS with supernatants of plasma-treated cancer cells may argue for the activation of some THP-1 monocytes. Along similar lines, the subtle but consistent decrease in metabolic activity is associated with activation of monocytes [61]. The levels of IL8 in ovarian cancer supernatants were reasonably high and followed the pattern of ROS in THP-1 monocytes. As IL8 is known to activate monocytes [62], it is reasonable to speculate on its prime role in mediating immediate changes. The decrease of IL8 in plasma-treated supernatants may also explain the alleviated morphometric responses observed with control tumor cell supernatants. A similar observation was made for tumor-derived small particles such as microparticles. These are known for their immunomodulatory properties [63]. However, their quality (protein or mRNA cargo) and not necessarily quantity governs their physiological impact on other cell types. Another secretory product found in ovarian cancer cells was HSP27 but its release pattern does not correlate to any of the findings in THP-1 monocytes, although it is crucial in macrophage polarization [64]. However, we do not propose finding fully differentiated macrophages in our study but rather propose a principle immunomodulatory function of ovarian cancer cell secretory products in monocytes that can be targeted with plasma treatment. The practical implications of these findings are limited but it could be speculated that plasma treatment of micro metastasis within resection margins of ovarian cancer surgery may decrease the tumor cells' ability to educate monocyte for cancer-promoting purposes.

Both OVCAR-3 and SKOV-3 are capable of undergoing caspase-dependent apoptosis [65,66]. OVCAR-3 and SKOV-3 cells differentially responded to plasma treatment. While a decrease in metabolic activity and an increase in terminal cell death was present in both lines, apoptosis was only significantly enhanced in plasma-treated SKOV-3 but not OVCAR-3. However, this was due to a high, non-eliminable background cell death when detaching and suspending OVCAR-3 cells. Despite this, plasma treatment was still perceived and translated by OVCAR-3 cells into a biological response, as seen with reduction in metabolic activity, release of HSP27, and modulation of cytokine secretion. In general, both OVCAR-3 and SKOV-3 have permanently active hypoxia-related redox-signaling pathways activated to respond to e.g., plasma-derived oxidants, as high steady-state expression levels of HIF1α suggests [67]. Although SKOV-3 viability was compromised to a higher extent compared to that of OVCAR-3, the cells' sensitivity towards ionizing radiation-induced cell death is reciprocal to our observation to plasma [68]. In addition, their cell surface marker expression differs to some extent, and SKOV-3 but not OVCAR-3 responds in a highly sensitive way towards platinum-based compounds [69]. Similar to plasma, such compounds are known to evoke oxidative stress [70]. For regular chemotherapeutic agents, SKOV-3 are more resistant compared to OVCAR-3 [71].

Author Contributions: Conceptualization, S.B.; Methodology, S.B. and E.F.; Formal Analysis, S.B., C.P.W., and E.F.; Investigation, C.P.W., and E.F.; Writing-Original Draft Preparation, S.B.; Writing-Review and Editing, D.K., A.M., K.-D.W., and M.B.S.; Supervision, S.B. and M.B.S.; Project Administration, S.B.; Funding Acquisition, S.B. and K.-D.W.

Funding: This research was funded by the German Federal Ministry of Education and Research, grant number 03Z22DN11.

Acknowledgments: The authors gratefully acknowledge technical support by Felix Niessner.

Conflicts of Interest: The authors declare no conflict of interest. The funders had no role in the design of the study; in the collection, analyses, or interpretation of data; in the writing of the manuscript, and in the decision to publish the results.

References

1. Weltmann, K.D.; von Woedtke, T. Plasma medicine—Current state of research and medical application. *Plasma Phys. Control. Fusion* **2017**, *1*, 014031. [CrossRef]
2. Laroussi, M. From killing bacteria to destroying cancer cells: 20 years of plasma medicine. *Plasma Process. Polym.* **2014**, *11*, 1138–1141. [CrossRef]
3. Guerrero-Preston, R.; Ogawa, T.; Uemura, M.; Shumulinsky, G.; Valle, B.L.; Pirini, F.; Ravi, R.; Sidransky, D.; Keidar, M.; Trink, B. Cold atmospheric plasma treatment selectively targets head and neck squamous cell carcinoma cells. *Int. J. Mol. Med.* **2014**, *34*, 941–946. [CrossRef] [PubMed]
4. Kim, S.Y.; Kim, H.J.; Kang, S.U.; Kim, Y.E.; Park, J.K.; Shin, Y.S.; Kim, Y.S.; Lee, K.; Kim, C.H. Non-thermal plasma induces AKT degradation through turn-on the MUL1 E3 ligase in head and neck cancer. *Oncotarget* **2015**, *6*, 33382–33396. [CrossRef] [PubMed]
5. Metelmann, H.-R.; Seebauer, C.; Miller, V.; Fridman, A.; Bauer, G.; Graves, D.B.; Pouvesle, J.-M.; Rutkowski, R.; Schuster, M.; Bekeschus, S.; et al. Clinical experience with cold plasma in the treatment of locally advanced head and neck cancer. *Clin. Plasma Med.* **2018**, *9*, 6–13. [CrossRef]
6. Turrini, E.; Laurita, R.; Stancampiano, A.; Catanzaro, E.; Calcabrini, C.; Maffei, F.; Gherardi, M.; Colombo, V.; Fimognari, C. Cold atmospheric plasma induces apoptosis and oxidative stress pathway regulation in t-lymphoblastoid leukemia cells. *Oxid. Med. Cell. Longev.* **2017**. [CrossRef] [PubMed]
7. Wang, C.; Zhang, H.X.; Xue, Z.X.; Yin, H.J.; Niu, Q.; Chen, H.L. The relation between doses or post-plasma time points and apoptosis of leukemia cells induced by dielectric barrier discharge plasma. *AIP Adv.* **2015**. [CrossRef]

8. Schmidt, A.; Rodder, K.; Hasse, S.; Masur, K.; Toups, L.; Lillig, C.H.; von Woedtke, T.; Wende, K.; Bekeschus, S. Redox-regulation of activator protein 1 family members in blood cancer cell lines exposed to cold physical plasma-treated medium. *Plasma Process. Polym.* **2016**, *13*, 1179–1188. [CrossRef]

9. Vandamme, M.; Robert, E.; Lerondel, S.; Sarron, V.; Ries, D.; Dozias, S.; Sobilo, J.; Gosset, D.; Kieda, C.; Legrain, B.; et al. Ros implication in a new antitumor strategy based on non-thermal plasma. *Int. J. Cancer* **2012**, *130*, 2185–2194. [CrossRef] [PubMed]

10. Tanaka, H.; Mizuno, M.; Ishikawa, K.; Nakamura, K.; Kajiyama, H.; Kano, H.; Kikkawa, F.; Hori, M. Plasma-activated medium selectively kills glioblastoma brain tumor cells by down-regulating a survival signaling molecule, AKT kinase. *Plasma Med.* **2011**, *1*, 265–277. [CrossRef]

11. Koritzer, J.; Boxhammer, V.; Schafer, A.; Shimizu, T.; Klampfl, T.G.; Li, Y.F.; Welz, C.; Schwenk-Zieger, S.; Morfill, G.E.; Zimmermann, J.L.; et al. Restoration of sensitivity in chemo-resistant glioma cells by cold atmospheric plasma. *PLoS ONE* **2013**. [CrossRef] [PubMed]

12. Brulle, L.; Vandamme, M.; Ries, D.; Martel, E.; Robert, E.; Lerondel, S.; Trichet, V.; Richard, S.; Pouvesle, J.M.; Le Pape, A. Effects of a non thermal plasma treatment alone or in combination with gemcitabine in a MIA PaCa2-luc orthotopic pancreatic carcinoma model. *PLoS ONE* **2012**. [CrossRef] [PubMed]

13. Hattori, N.; Yamada, S.; Torii, K.; Takeda, S.; Nakamura, K.; Tanaka, H.; Kajiyama, H.; Kanda, M.; Fujii, T.; Nakayama, G.; et al. Effectiveness of plasma treatment on pancreatic cancer cells. *Int. J. Oncol.* **2015**, *47*, 1655–1662. [CrossRef] [PubMed]

14. Liedtke, K.R.; Bekeschus, S.; Kaeding, A.; Hackbarth, C.; Kuehn, J.P.; Heidecke, C.D.; von Bernstorff, W.; von Woedtke, T.; Partecke, L.I. Non-thermal plasma-treated solution demonstrates antitumor activity against pancreatic cancer cells in vitro and in vivo. *Sci. Rep.* **2017**. [CrossRef] [PubMed]

15. Bekeschus, S.; Rodder, K.; Fregin, B.; Otto, O.; Lippert, M.; Weltmann, K.D.; Wende, K.; Schmidt, A.; Gandhirajan, R.K. Toxicity and immunogenicity in murine melanoma following exposure to physical plasma-derived oxidants. *Oxid. Med. Cell. Longev.* **2017**. [CrossRef] [PubMed]

16. Ishaq, M.; Kumar, S.; Varinli, H.; Han, Z.J.; Rider, A.E.; Evans, M.D.; Murphy, A.B.; Ostrikov, K. Atmospheric gas plasma-induced ros production activates TNF-ASK1 pathway for the induction of melanoma cancer cell apoptosis. *Mol. Biol. Cell* **2014**, *25*, 1523–1531. [CrossRef] [PubMed]

17. Mizuno, K.; Shirakawa, Y.; Sakamoto, T.; Ishizaki, H.; Nishijima, Y.; Ono, R. Plasma-induced suppression of recurrent and reinoculated melanoma tumors in mice. *IEEE Trans. Radiat. Plasma Med. Sci.* **2018**, *2*, 353–359. [CrossRef]

18. Lin, A.G.; Xiang, B.; Merlino, D.J.; Baybutt, T.R.; Sahu, J.; Fridman, A.; Snook, A.E.; Miller, V. Non-thermal plasma induces immunogenic cell death in vivo in murine CT26 colorectal tumors. *OncoImmunology* **2018**, 1–13. [CrossRef]

19. Bekeschus, S.; Mueller, A.; Miller, V.; Gaipl, U.; Weltmann, K.-D. Physical plasma elicits immunogenic cancer cell death and mitochondrial singlet oxygen. *IEEE Trans. Radiat. Plasma Med. Sci.* **2018**, *2*, 138–146. [CrossRef]

20. Plewa, J.M.; Yousfi, M.; Frongia, C.; Eichwald, O.; Ducommun, B.; Merbahi, N.; Lobjois, V. Low-temperature plasma-induced antiproliferative effects on multi-cellular tumor spheroids. *New J. Phys.* **2014**. [CrossRef]

21. Hirst, A.M.; Simms, M.S.; Mann, V.M.; Maitland, N.J.; O'Connell, D.; Frame, F.M. Low-temperature plasma treatment induces DNA damage leading to necrotic cell death in primary prostate epithelial cells. *Br. J. Cancer* **2015**, *112*, 1536–1545. [CrossRef] [PubMed]

22. Weiss, M.; Gumbel, D.; Hanschmann, E.M.; Mandelkow, R.; Gelbrich, N.; Zimmermann, U.; Walther, R.; Ekkernkamp, A.; Sckell, A.; Kramer, A.; et al. Cold atmospheric plasma treatment induces anti-proliferative effects in prostate cancer cells by redox and apoptotic signaling pathways. *PLoS ONE* **2015**. [CrossRef] [PubMed]

23. Zhunussova, A.; Vitol, E.A.; Polyak, B.; Tuleukhanov, S.; Brooks, A.D.; Sensenig, R.; Friedman, G.; Orynbayeva, Z. Mitochondria-mediated anticancer effects of non-thermal atmospheric plasma. *PLoS ONE* **2016**. [CrossRef] [PubMed]

24. Gumbel, D.; Gelbrich, N.; Napp, M.; Daeschlein, G.; Kramer, A.; Sckell, A.; Burchardt, M.; Ekkernkamp, A.; Stope, M.B. Peroxiredoxin expression of human osteosarcoma cells is influenced by cold atmospheric plasma treatment. *Anticancer Res.* **2017**, *37*, 1031–1038. [PubMed]

25. Tokunaga, T.; Ando, T.; Suzuki-Karasaki, M.; Ito, T.; Onoe-Takahashi, A.; Ochiai, T.; Soma, M.; Suzuki-Karasaki, Y. Plasma-stimulated medium kills trail-resistant human malignant cells by promoting caspase-independent cell death via membrane potential and calcium dynamics modulation. *Int. J. Oncol.* **2018**, *52*, 697–708. [CrossRef] [PubMed]

26. Canal, C.; Fontelo, R.; Hamouda, I.; Guillem-Marti, J.; Cvelbar, U.; Ginebra, M.P. Plasma-induced selectivity in bone cancer cells death. *Free Radic. Biol. Med.* **2017**, *110*, 72–80. [CrossRef] [PubMed]

27. Koensgen, D.; Besic, I.; Gumbel, D.; Kaul, A.; Weiss, M.; Diesing, K.; Kramer, A.; Bekeschus, S.; Mustea, A.; Stope, M.B. Cold atmospheric plasma (CAP) and CAP-stimulated cell culture media suppress ovarian cancer cell growth—A putative treatment option in ovarian cancer therapy. *Anticancer Res.* **2017**, *37*, 6739–6744. [PubMed]

28. Nakamura, K.; Peng, Y.; Utsumi, F.; Tanaka, H.; Mizuno, M.; Toyokuni, S.; Hori, M.; Kikkawa, F.; Kajiyama, H. Novel intraperitoneal treatment with non-thermal plasma-activated medium inhibits metastatic potential of ovarian cancer cells. *Sci. Rep.* **2017**. [CrossRef] [PubMed]

29. Utsumi, F.; Kajiyama, H.; Nakamura, K.; Tanaka, H.; Hori, M.; Kikkawa, F. Selective cytotoxicity of indirect nonequilibrium atmospheric pressure plasma against ovarian clear-cell carcinoma. *Springerplus* **2014**. [CrossRef] [PubMed]

30. Ahn, H.J.; Kim, K.I.; Hoan, N.N.; Kim, C.H.; Moon, E.; Choi, K.S.; Yang, S.S.; Lee, J.S. Targeting cancer cells with reactive oxygen and nitrogen species generated by atmospheric-pressure air plasma. *PLoS ONE* **2014**. [CrossRef] [PubMed]

31. Graves, D.B. The emerging role of reactive oxygen and nitrogen species in redox biology and some implications for plasma applications to medicine and biology. *J. Phys. D Appl. Phys.* **2012**. [CrossRef]

32. Yan, D.; Sherman, J.H.; Keidar, M. Cold atmospheric plasma, a novel promising anti-cancer treatment modality. *Oncotarget* **2017**, *8*, 15977–15995. [CrossRef] [PubMed]

33. Wang, H.; Naghavi, M.; Allen, C.; Barber, R.M.; Bhutta, Z.A.; Carter, A.; Casey, D.C.; Charlson, F.J.; Chen, A.Z.; Coates, M.M.; et al. Global, regional, and national life expectancy, all-cause mortality, and cause-specific mortality for 249 causes of death, 1980–2015: A systematic analysis for the global burden of disease study 2015. *Lancet* **2016**, *388*, 1459–1544. [CrossRef]

34. Wang, X.; Deavers, M.; Patenia, R.; Bassett, R.L., Jr.; Mueller, P.; Ma, Q.; Wang, E.; Freedman, R.S. Monocyte/macrophage and T-cell infiltrates in peritoneum of patients with ovarian cancer or benign pelvic disease. *J. Transl. Med.* **2006**. [CrossRef]

35. Kawamura, K.; Komohara, Y.; Takaishi, K.; Katabuchi, H.; Takeya, M. Detection of M2 macrophages and colony-stimulating factor 1 expression in serous and mucinous ovarian epithelial tumors. *Pathol. Int.* **2009**, *59*, 300–305. [CrossRef] [PubMed]

36. Sica, A.; Larghi, P.; Mancino, A.; Rubino, L.; Porta, C.; Totaro, M.G.; Rimoldi, M.; Biswas, S.K.; Allavena, P.; Mantovani, A. Macrophage polarization in tumour progression. *Semin. Cancer Biol.* **2008**, *18*, 349–355. [CrossRef] [PubMed]

37. Takaishi, K.; Komohara, Y.; Tashiro, H.; Ohtake, H.; Nakagawa, T.; Katabuchi, H.; Takeya, M. Involvement of M2-polarized macrophages in the ascites from advanced epithelial ovarian carcinoma in tumor progression via Stat3 activation. *Cancer Sci.* **2010**, *101*, 2128–2136. [CrossRef] [PubMed]

38. Henze, A.T.; Mazzone, M. The impact of hypoxia on tumor-associated macrophages. *J. Clin. Invest.* **2016**, *126*, 3672–3679. [CrossRef] [PubMed]

39. Sica, A.; Schioppa, T.; Mantovani, A.; Allavena, P. Tumour-associated macrophages are a distinct M2 polarised population promoting tumour progression: Potential targets of anti-cancer therapy. *Eur. J. Cancer* **2006**, *42*, 717–727. [CrossRef] [PubMed]

40. Freedman, R.S.; Deavers, M.; Liu, J.; Wang, E. Peritoneal inflammation—A microenvironment for epithelial ovarian cancer (EOC). *J. Transl. Med.* **2004**. [CrossRef] [PubMed]

41. Quail, D.F.; Joyce, J.A. Microenvironmental regulation of tumor progression and metastasis. *Nat. Med.* **2013**, *19*, 1423–1437. [CrossRef] [PubMed]

42. Coussens, L.M.; Werb, Z. Inflammation and cancer. *Nature* **2002**, *420*, 860–867. [CrossRef] [PubMed]

43. Bekeschus, S.; Moritz, J.; Schmidt, A.; Wende, K. Redox regulation of leukocyte-derived microparticle release and protein content in response to cold physical plasma-derived oxidants. *Clin. Plasma Med.* **2017**. [CrossRef]

44. Baj-Krzyworzeka, M.; Mytar, B.; Szatanek, R.; Surmiak, M.; Weglarczyk, K.; Baran, J.; Siedlar, M. Colorectal cancer-derived microvesicles modulate differentiation of human monocytes to macrophages. *J. Transl. Med.* **2016**. [CrossRef] [PubMed]

45. Salari, S.; Seibert, T.; Chen, Y.X.; Hu, T.; Shi, C.; Zhao, X.; Cuerrier, C.M.; Raizman, J.E.; O'Brien, E.R. Extracellular HSP27 acts as a signaling molecule to activate NF-κB in macrophages. *Cell Stress Chaperones* **2013**, *18*, 53–63. [CrossRef] [PubMed]

46. Zhou, R.; Yazdi, A.S.; Menu, P.; Tschopp, J. A role for mitochondria in NLRP3 inflammasome activation. *Nature* **2011**, *469*, 221–225. [CrossRef] [PubMed]

47. Auwerx, J. The human leukemia cell line, THP-1: A multifacetted model for the study of monocyte-macrophage differentiation. *Experientia* **1991**, *47*, 22–31. [CrossRef] [PubMed]

48. He, Q.; You, H.; Li, X.-M.; Liu, T.-H.; Wang, P.; Wang, B.-E. HMGB1 promotes the synthesis of pro-IL-1β and pro-IL-18 by activation of p38 MAPK and NF-κB through receptors for advanced glycation end-products in macrophages. *Asian Pac. J. Cancer Prev.* **2012**, *13*, 1365–1370. [CrossRef] [PubMed]

49. Sonoda, Y.; Kasahara, T.; Mukaida, N.; Shimizu, N.; Tomoda, M.; Takeda, T. Stimulation of interleukin-8 production by acidic polysaccharides from the root of panax ginseng. *Immunopharmacology* **1998**, *38*, 287–294. [CrossRef]

50. Zhang, M.; Zhao, G.J.; Yin, K.; Xia, X.D.; Gong, D.; Zhao, Z.W.; Chen, L.Y.; Zheng, X.L.; Tang, X.E.; Tang, C.K. Apolipoprotein A-1 binding protein inhibits inflammatory signaling pathways by binding to apolipoprotein A-1 in THP-1 macrophages. *Circ. J.* **2018**, *82*, 1396–1404. [CrossRef] [PubMed]

51. Bekeschus, S.; Schmidt, A.; Bethge, L.; Masur, K.; von Woedtke, T.; Hasse, S.; Wende, K. Redox stimulation of human THP-1 monocytes in response to cold physical plasma. *Oxid. Med. Cell. Longev.* **2016**. [CrossRef] [PubMed]

52. Daigneault, M.; Preston, J.A.; Marriott, H.M.; Whyte, M.K.; Dockrell, D.H. The identification of markers of macrophage differentiation in PMA-stimulated THP-1 cells and monocyte-derived macrophages. *PLoS ONE* **2010**. [CrossRef] [PubMed]

53. Sozzani, S.; Molino, M.; Locati, M.; Luini, W.; Cerletti, C.; Vecchi, A.; Mantovani, A. Receptor-activated calcium influx in human monocytes exposed to monocyte chemotactic protein-1 and related cytokines. *J. Immunol.* **1993**, *150*, 1544–1553. [PubMed]

54. Jablonowski, H.; von Woedtke, T. Research on plasma medicine-relevant plasma–liquid interaction: What happened in the past five years? *Clin. Plasma Med.* **2015**, *3*, 42–52. [CrossRef]

55. Bekeschus, S.; Kolata, J.; Winterbourn, C.; Kramer, A.; Turner, R.; Weltmann, K.D.; Broker, B.; Masur, K. Hydrogen peroxide: A central player in physical plasma-induced oxidative stress in human blood cells. *Free Radic. Res.* **2014**, *48*, 542–549. [CrossRef] [PubMed]

56. Bekeschus, S.; Winterbourn, C.C.; Kolata, J.; Masur, K.; Hasse, S.; Broker, B.M.; Parker, H.A. Neutrophil extracellular trap formation is elicited in response to cold physical plasma. *J. Leukoc. Biol.* **2016**, *100*, 791–799. [CrossRef] [PubMed]

57. Pelicano, H.; Carney, D.; Huang, P. Ros stress in cancer cells and therapeutic implications. *Drug Resist. Updat.* **2004**, *7*, 97–110. [CrossRef] [PubMed]

58. Cai, J.; Jones, D.P. Superoxide in apoptosis. Mitochondrial generation triggered by cytochrome c loss. *J. Biol. Chem.* **1998**, *273*, 11401–11404. [CrossRef] [PubMed]

59. Lackmann, J.W.; Wende, K.; Verlackt, C.; Golda, J.; Volzke, J.; Kogelheide, F.; Held, J.; Bekeschus, S.; Bogaerts, A.; Schulz-von der Gathen, V.; et al. Chemical fingerprints of cold physical plasmas—An experimental and computational study using cysteine as tracer compound. *Sci. Rep.* **2018**. [CrossRef] [PubMed]

60. Kasai, S.; Shiku, H.; Torisawa, Y.-s.; Noda, H.; Yoshitake, J.; Shiraishi, T.; Yasukawa, T.; Watanabe, T.; Matsue, T.; Yoshimura, T. Real-time monitoring of reactive oxygen species production during differentiation of human monocytic cell lines (THP-1). *Anal. Chim. Acta* **2005**, *549*, 14–19. [CrossRef]

61. Traore, K.; Trush, M.A.; George, M., Jr.; Spannhake, E.W.; Anderson, W.; Asseffa, A. Signal transduction of phorbol 12–myristate 13-acetate (PMA)-induced growth inhibition of human monocytic leukemia THP-1 cells is reactive oxygen dependent. *Leuk. Res.* **2005**, *29*, 863–879. [CrossRef] [PubMed]

62. Gerszten, R.E.; Garcia-Zepeda, E.A.; Lim, Y.C.; Yoshida, M.; Ding, H.A.; Gimbrone, M.A., Jr.; Luster, A.D.; Luscinskas, F.W.; Rosenzweig, A. MCP-1 and IL-8 trigger firm adhesion of monocytes to vascular endothelium under flow conditions. *Nature* **1999**, *398*, 718–723. [CrossRef] [PubMed]

63. Camussi, G.; Deregibus, M.C.; Tetta, C. Tumor-derived microvesicles and the cancer microenvironment. *Curr. Mol. Med.* **2013**, *13*, 58–67. [CrossRef] [PubMed]

64. Fagone, P.; Di Rosa, M.; Palumbo, M.; De Gregorio, C.; Nicoletti, F.; Malaguarnera, L. Modulation of heat shock proteins during macrophage differentiation. *Inflamm. Res.* **2012**, *61*, 1131–1139. [CrossRef] [PubMed]

65. Olichon, A.; Baricault, L.; Gas, N.; Guillou, E.; Valette, A.; Belenguer, P.; Lenaers, G. Loss of OPA1 perturbates the mitochondrial inner membrane structure and integrity, leading to cytochrome c release and apoptosis. *J. Biol. Chem.* **2003**, *278*, 7743–7746. [CrossRef] [PubMed]

66. Yeo, J.-K.; Cha, S.-D.; Cho, C.-H.; Kim, S.-P.; Cho, J.-W.; Baek, W.-K.; Suh, M.-H.; Kwon, T.K.; Park, J.-W.; Suh, S.-I. Se-methylselenocysteine induces apoptosis through caspase activation and bax cleavage mediated by calpain in SKOV-3 ovarian cancer cells. *Cancer Lett.* **2002**, *182*, 83–92. [CrossRef]

67. Skinner, H.D.; Zheng, J.Z.; Fang, J.; Agani, F.; Jiang, B.H. Vascular endothelial growth factor transcriptional activation is mediated by hypoxia-inducible factor 1alpha, HDM2, and p70S6K1 in response to phosphatidylinositol 3-kinase/AKT signaling. *J. Biol. Chem.* **2004**, *279*, 45643–45651. [CrossRef] [PubMed]

68. Palm, S.; Back, T.; Claesson, I.; Danielsson, A.; Elgqvist, J.; Frost, S.; Hultborn, R.; Jensen, H.; Lindegren, S.; Jacobsson, L. Therapeutic efficacy of astatine-211-labeled trastuzumab on radioresistant SKOV-3 tumors in nude mice. *Int. J. Radiat. Oncol. Biol. Phys.* **2007**, *69*, 572–579. [CrossRef] [PubMed]

69. Hills, C.; Kelland, L.; Abel, G.; Siracky, J.; Wilson, A.; Harrap, K. Biological properties of ten human ovarian carcinoma cell lines: Calibration in vitro against four platinum complexes. *Br. J. Cancer* **1989**, *59*, 527. [CrossRef] [PubMed]

70. Olas, B.; Wachowicz, B.; Majsterek, I.; Blasiak, J. Resveratrol may reduce oxidative stress induced by platinum compounds in human plasma, blood platelets and lymphocytes. *Anticancer Drugs* **2005**, *16*, 659–665. [CrossRef] [PubMed]

71. Petru, E.; Sevin, B.U.; Perras, J.; Boike, G.; Ramos, R.; Nguyen, H.; Averette, H.E. Comparative chemosensitivity profiles in four human ovarian carcinoma cell lines measuring ATP bioluminescence. *Gynecol. Oncol.* **1990**, *38*, 155–160. [CrossRef]

Article

The Canady Helios Cold Plasma Scalpel Significantly Decreases Viability in Malignant Solid Tumor Cells in a Dose-Dependent Manner

Warren Rowe [1],[†] , Xiaoqian Cheng [1],[†], Lawan Ly [1], Taisen Zhuang [2], Giacomo Basadonna [1],[3], Barry Trink [1],[4], Michael Keidar [1],[4] and Jerome Canady [1],[4],*

[1]   Jerome Canady Research Institute for Advanced Biological and Technological Sciences,
       Takoma Park, MD 20912, USA; drwrowe@usmedinnov.com (W.R.); xcheng@usmedinnov.com (X.C.);
       llawan@usmedinnov.com (L.L.); Giacomo.basadonna@umassmed.edu (G.B.); barrytrink@gmail.com (B.T.);
       keidar@gwu.edu (M.K.)
[2]   Plasma Medicine Life Sciences, Takoma Park, MD 20912, USA; tzhuang@usmedinnov.com
[3]   Department of Surgery, University of Massachusetts School of Medicine, Worcester, MA 01655, USA
[4]   School of Engineering & Applied Science, The George Washington University, Washington, DC 20052, USA
*   Correspondence: drjcanady@usmedinnov.com; Tel.: +301-270-0147
†   The authors contributed equally to this work.

Received: 19 July 2018; Accepted: 24 August 2018; Published: 7 September 2018

Abstract: To determine appropriate treatment doses of cold atmospheric plasma (CAP), the Canady Helios Cold Plasma Scalpel was tested across numerous cancer cell types including renal adenocarcinoma, colorectal carcinoma, pancreatic adenocarcinoma, ovarian adenocarcinoma, and esophageal adenocarcinoma. Various CAP doses were tested consisting of both high (3 L/min) and low (1 L/min) helium flow rates, several power settings, and a range of treatment times up to 5 min. The impact of cold plasma on the reduction of viability was consistently dose-dependent; however, the anti-cancer capability varied significantly between cell lines. While the lowest effective dose varied from cell line to cell line, in each case an 80–99% reduction in viability was achievable 48 h after CAP treatment. Therefore, it is critical to select the appropriate CAP dose necessary for treating a specific cancer cell type.

Keywords: cold atmospheric plasma; CAP; cancer therapy; dose-dependent; renal adenocarcinoma; colorectal carcinoma; pancreatic adenocarcinoma; ovarian adenocarcinoma; esophageal adenocarcinoma

---

1. Introduction

Malignant solid tumors are characterized by high recurrence rates and low five-year survival rates. Stage IV renal adenocarcinoma presents an extremely low five-year survival rate of 0–10% [1], while the recurrence rate may be as high as 23% [2]. While the recurrence rate of colorectal carcinoma is similar to that of renal adenocarcinoma at 19.4% to 21.6%, the five-year survival rate is significantly higher at 88.6% to 89.4% [3]. Pancreatic ductal adenocarcinoma has an extremely low survival rate of 10% to 28% [4] after one year due to a very high recurrence rate of 65.5% [5]. The resulting 5-year survival rate is dismal at 6% in the United States and Europe [6]. Serous epithelial ovarian carcinoma has a very low five-year survival rate of 42% for stage III and 26% for stage IV [7] with a 19% recurrence rate [8]. Esophageal adenocarcinoma represents a similarly low five-year survival rate of 33% to 44% [9], depending on the treatment used, and can result in a recurrence rate as high as 43.2% [10].

The unfortunately common recurrence of malignant solid tumors represents a unique opportunity for cold atmospheric plasma (CAP) treatment. CAP is an emerging technology that uses ionized gas to produce a plasma beam which has numerous applications in various fields, including dentistry, wound healing, surface decontamination, viral inhibition, and cancer treatment [11–16]. CAP can be used to treat the margins following tumor removal, and in doing so has the potential to remove residual cancer cells and prevent recurrence. An important step to make this a reality is to determine the correct dose of CAP to significantly reduce tumor cell viability. It has been reported that various cell lines react differently to CAP treatment [17–20]. Yan et al. studied the reactive species consumption speed of glioblastoma U87 and breast cancers MDA-MB-231 and MCF-7, and discovered that the cancer cells that could absorb or eliminate the effective species in the media faster (glioblastoma) were more resistant to plasma-activated media than both breast cancers [17]. Their results also demonstrate a wide range of effects on cell viability depending on cell type and treatment time [20]. Naciri et al. also reported that plasma sensitivity closely correlates with proliferation rates by measuring ATP levels of three cancer cell types including Chinese hamster ovary cells, osteoblast, and colon adenocarcinoma [18]. By testing eight cancerous cell lines, Ma et al. claimed that p53-deficient cancer cells are more sensitive to CAP treatment due to the lack of p53-dependent cell cycle delay at G1 [19]. Therefore, the combinations of power settings and treatment times are critical to establish the correct dosage of cell line-dependent CAP treatment.

The mechanism of the anti-cancer capacity of CAP has been increasingly understood. Several theories have been proposed, including a decrease of cell adhesion [21,22], interruption of the cell cycle [23–25], induction of apoptosis [26–29], and DNA fragmentation [30]. However, it is not yet clear whether CAP-generated reactive species are crucial for apoptosis and its associated DNA strand break or whether plasma-induced direct DNA damage provokes cell cycle checkpoint signaling that leads to apoptosis [31].

The Canady Cold Plasma Conversion Unit is unique in that it utilizes a high voltage transformer to up-convert the voltage (1.5–50 kV), down-convert the frequency (<300 kHz), and down-convert the power (<30 W) of the high-voltage output from an electrosurgical unit (U.S. Patent No. 9,999,462) [32]. We tested the Canady Helios Cold Plasma Device on a wide range of cell lines with different combinations of power settings, treatment times, and gas flow rates. With optimal dosage for each cancer type, this study provides a starting point for future animal studies and clinical trials.

2. Materials and Methods

2.1. Cold Plasma Device

All experiments were performed at the Jerome Canady Research Institute for Advanced Biological and Technological Sciences, in Takoma Park, MD, USA. Cold atmospheric plasma (CAP) was generated using a USMI SS-601 MCa high-frequency electrosurgical generator (USMI, Takoma Park, MD, USA) integrated with a USMI Cold Plasma Conversion Unit and connected to a Canady Helios Cold Plasma Scalpel. Helium flow was set to a constant 1 L/min and 20 P or 40 P or 3 L/min and power set to 40 P, 60 P, or 80 P. The plasma scalpel was placed so that the tip of the scalpel was 1.5 cm (at 1 L/min) or 2 cm (at 3 L/min) from the surface of the cell media and was not moved during treatment (Figure 1).

2.2. Optical Emission Spectroscopy

Optical emission spectroscopy with a range of 250–850 nm was performed on the CAP jet to detect the reactive species. The spectrometer (EPP2000-HR) and detection probe were purchased from Stellar Net Inc. (Tampa, FL, USA). The probe was placed 1.5 cm away from the tip of the scalpel, perpendicular to the plasma beam. The integration time was set to 100 ms.

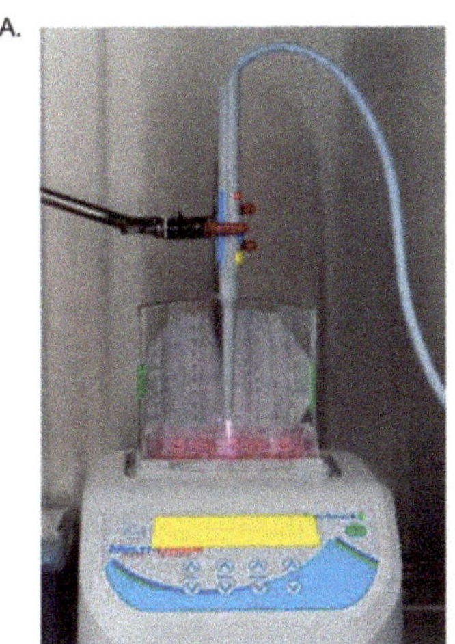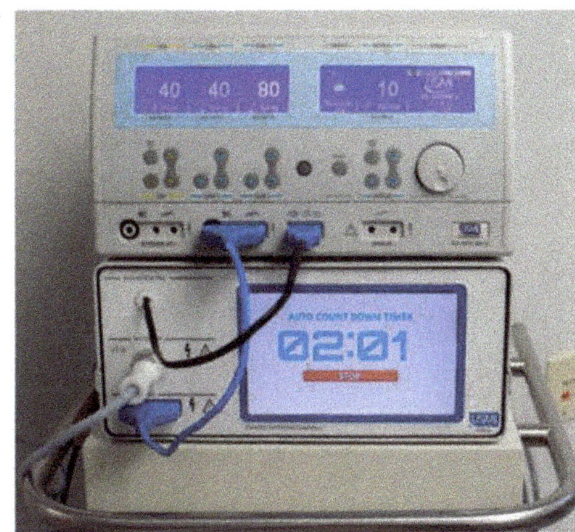

Figure 1. Solid tumor treatment with cold atmospheric plasma. (**A**) Cells cultured in a 12-well plate, treated using a Canady Helios Plasma Scalpel (USMI, Takoma Park, MD, USA) at 3 L/min (**B**) Cold atmospheric plasma (CAP) generator device; USMI SS-601 MCa (top) connected to a USMI Cold Plasma Conversion Unit (bottom).

2.3. Cell Culture

BxPC-3 pancreatic adenocarcinoma and 769-P renal adenocarcinoma were purchased from ATCC (Manassas, VA, USA). OE33 esophageal adenocarcinoma was purchased from Sigma-Aldrich (St. Louis, MO, USA). HCT-116 colorectal carcinoma and SK-OV-3 ovarian adenocarcinoma were generously donated by Professor Keidar's lab at The George Washington University. All cell lines were maintained with the required culture media according to the supplier protocol. When cells reached approximately 80% confluence, they were seeded at a concentration of 5×10^3 or 10^5 cells/well into 96-well or 12-well plates (USA Scientific, Ocala, FL, USA), respectively, for cell viability assays. For BxPC-3, 1×10^4 cells were required for the 96-well assay.

2.4. Cell Viability Assay

Thiazolyl blue tetrazolium bromide (MTT) assay was performed on the cells 48 h after plasma treatment following the manufacturer's protocol. Briefly, cells were incubated with MTT at a concentration of 0.5 mg/mL 48 h post treatment for 3 h in a 37 °C and 5% CO_2 humidified incubator. Then, MTT solvent was added into each well to dissolve the formazan crystals. All the MTT assay reagents were purchased from Sigma-Aldrich (St. Louis, MO, USA). The absorbance of the dissolved compound was measured by BioTek Synergy HTX (Winooski, VT, USA) microplate reader at 570 nm. Cell viability assay data has been included as supplementary material.

2.5. Statistics

All viability assays were repeated at least 3 times with 2 replicates each. Data was plotted by Microsoft Excel 2016 as the mean ± standard error of the mean. A student *t*-test or a one-way analysis of variance (ANOVA) was used to check statistical significance where applicable. The differences were considered statistically significant for * $p \leq 0.05$.

3. Results

Reduction of Cell Viability by CAP in Malignant Solid Tumor Cell Lines

To determine the effective plasma dose required to significantly reduce cell viability two flow rates were chosen; 1 L/min and 3 L/min of helium. Helium was used as the carrier gas because the breakdown voltage of helium is significantly lower than nitrogen [33]. In addition, the low excitation energy level of argon metastables (11.7 and 11.5 eV) in comparison with helium (20.6 and 19.8 eV) does not allow an efficient Penning ionization to sustain the discharge [34]. Based on initial testing, 1–5 min with 40–80 power, and 30–120 s with 20–40 power, were chosen as an effective range for 3 L/min and 1 L/min, respectively. The power settings of 20 P, 40 P, 60 P, and 80 P used in this study are 5 W, 8 W, 11 W, and 15.7 W at 3 L/min. At 1 L/min of 20 P and 40 P, the power settings are 5 W and 6 W respectively. The detailed power measurement of our CAP device was conducted and reported in another paper which is currently under review [35]. The spectrum of the CAP jet is shown in Figure 2. The most intense peaks and bands of plasma between 250 and 850 nm were referenced [36] and labeled in the figure. The main species observed were: OH ($A^2\Sigma^+$–$X^2\Pi^+$) at 309 nm, N_2 ($C^3\Pi_u$–$B^3\Pi_g$) second positive system (SPS) at 337 and 357 nm; N_2^+ ($B^2\Sigma_u^+$–$X^2\Sigma_g^+$) first negative system (FNS) at 391 and 427 nm; He at 667 nm; and O^I at 777 nm. The cold plasma jet is a complicated environment that combines the comprehensive effect of a variety of ions and neutrals. These reactive oxygen and nitrogen species are playing essential roles in cellular responses to the CAP treatment [37,38].

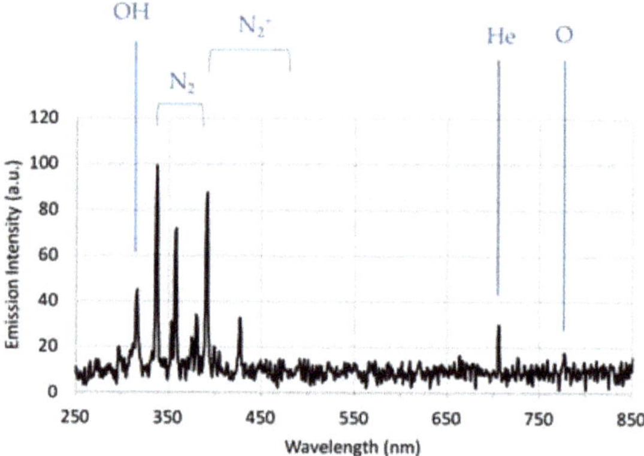

Figure 2. The spectrum of CAP generated by the USMI Cold Plasma Conversion Unit and Canady Helios Cold Plasma Scalpel. Data shown are in the optical emission spectroscopy within a range of 250–850 nm, measured 1.5 cm from the tip of the scalpel. Main species include OH (309 nm), N_2 (337 and 357 nm), N_2^+ (391 and 427 nm), He (667 nm), and O (777 nm).

MTT assays were used to determine the dose of CAP needed to significantly reduce cell viability. MTT assays were performed on CAP-treated cancer cell lines 48 h post treatment. The viability of the treated cells was normalized to an untreated (control) group. The viability of 769-P renal adenocarcinoma cells was dose-dependent and significantly reduced at all time and power combinations tested (Figure 3). Helium flow alone (0 W) did not significantly impact cell viability (Figure 3A). At the highest doses, using 3 L/min, viability was reduced to 4.9% ($p < 0.001$) while viability at 20 P for 120 s was <1% ($p < 0.001$) for 1 L/min. Increasing the power to 40 P did not result in a further reduction ($p = 0.65$) (Figure 3B). CAP was equally as effective in reducing viability in HCT-116 colorectal carcinoma cells. At 3 L/min, viability was 76% at 40 P for 1 min ($p < 0.01$) and this decreased to 3.6% at the highest dose of 80 P for 5 min ($p < 0.001$) (Figure 4A). The decrease in viability when using 1 L/min required a lower dose. Initially, at 20 P for 30 s and 60 s, viability was reduced to 27% ($p < 0.001$) and 21% ($p < 0.01$), respectively (Figure 4B). However, beginning at 20 P for 90 s viability was reduced to 5.0% ($p < 0.001$) which only slightly decreased with a higher dose, with the highest dose resulting in 2.6% viability ($p < 0.001$). CAP also had a clear dose-dependent effect on the reduction of viability in the ovarian adenocarcinoma cell line; SK-OV-3. At the lowest dose, using 3 L/min, viability was only reduced to 87% ($p < 0.001$), which decreased to 17% ($p < 0.001$) at 80 P for 5 min (Figure 5A). Similar results were found with a lower flow rate (Figure 5B). 20 P for 30 s resulted in 77% viability ($p < 0.001$), which decreased to 4% viability ($p < 0.0001$) at 40 P for 120 s. BxPC-3 (pancreatic adenocarcinoma) required a higher dose to effectively reduce viability (Figure 6A). At 60 P for 5 min and 80 P for 5 min, viability was reduced to 14% ($p < 0.0001$) and 4% ($p < 0.0001$), respectively. The low flow rate treatment showed a similar pattern requiring a higher dose to reduce viability. At 40 P for 90 s and 40 P for 120 s the viability was reduced to 5% ($p < 0.0001$) and 1% ($p < 0.0001$), respectively (Figure 6B). The esophageal adenocarcinoma cell line, OE33, also required a higher dose to decrease viability below 20%. Using the high flow rate at 80 P for 5 min viability was reduced to 16% ($p < 0.001$) (Figure 7A). At 1 L/min and 40 P for 120 s viability was reduced to 15% ($p < 0.0001$) (Figure 7B). Taken together, this data demonstrates that CAP reduced cell viability in a time- and power-dependent manner in all cell lines tested.

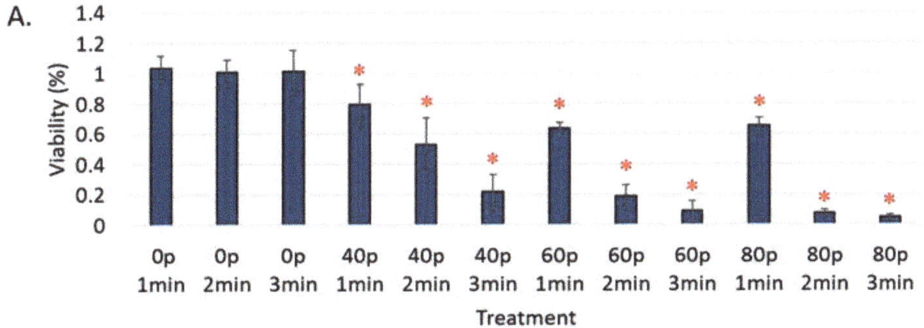

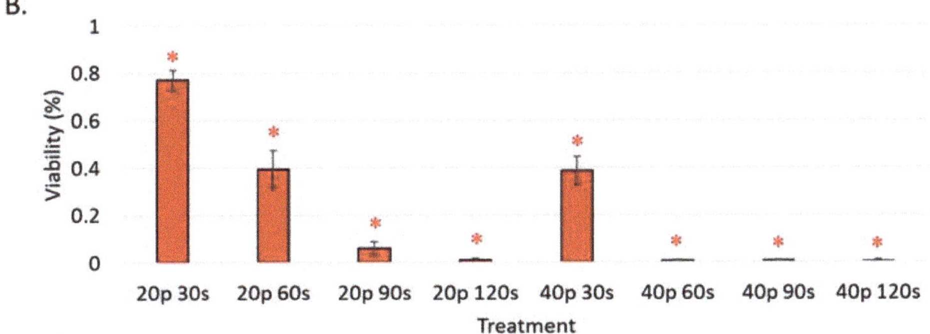

Figure 3. The reduction of viability of 769-P (renal adenocarcinoma) following CAP treatment. (**A**) 769-P cells were cultured in 12-well plates and treated with CAP at 3 L/min; (**B**) 769-P cells were cultured in 96-well plates and treated with CAP at 1 L/min. CAP treatment of 769-P significantly reduces viability at all doses tested, at both 3 L/min and 1 L/min. * $p \leq 0.05$.

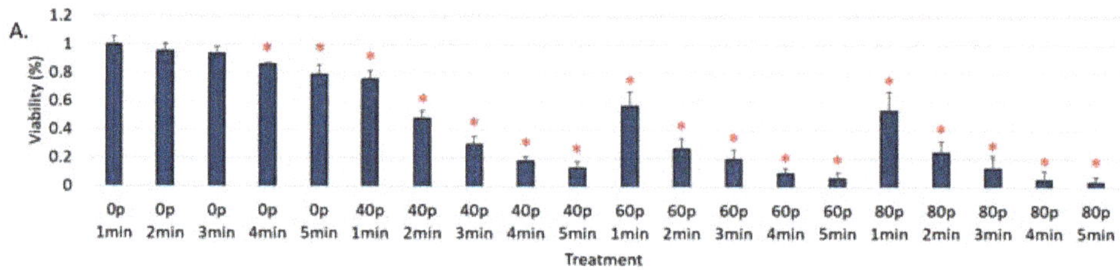

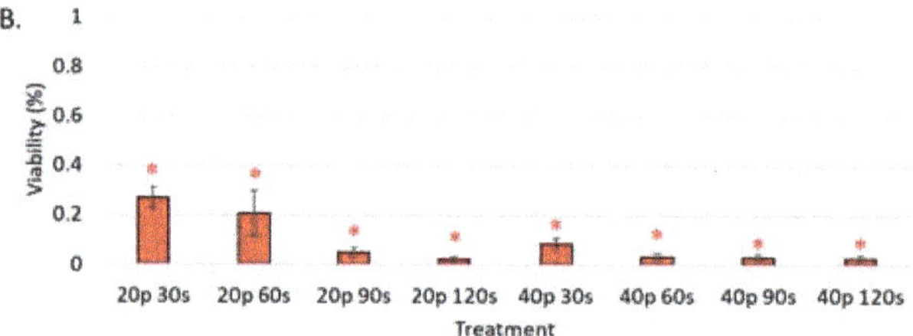

Figure 4. The reduction of the viability of HCT-116 (colorectal carcinoma) following CAP treatment. (**A**) HCT-116 cells were cultured in 12-well plates and treated with CAP at 3 L/min; (**B**) HCT-116 cells were cultured in 96-well plates and treated with CAP at 1 L/min. CAP treatment of HCT-116 significantly reduces viability at all doses tested, at both 3 L/min and 1 L/min. * $p \leq 0.05$.

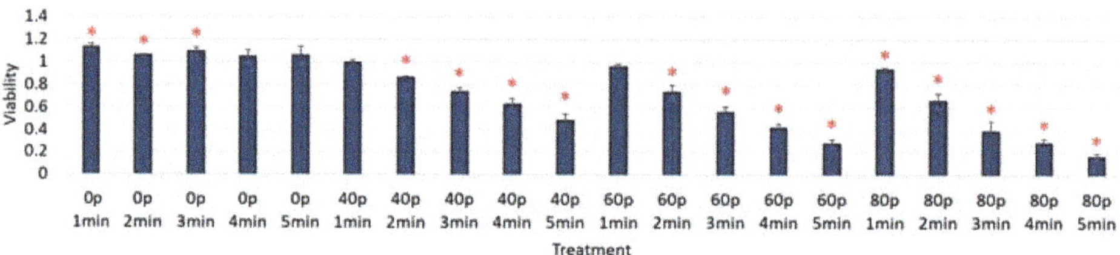

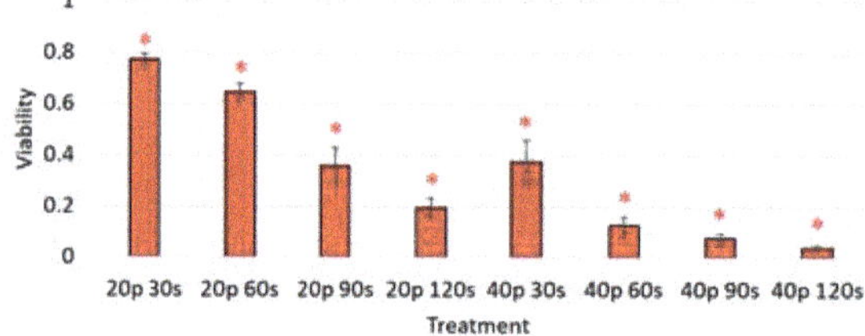

Figure 5. The reduction of the viability of SK-OV-3 (ovarian adenocarcinoma) following CAP treatment. (**top**) SK-OV-3 cells were cultured in 12-well plates and treated with CAP at 3 L/min; (**bottom**) SK-OV-3 cells were cultured in 96-well plates and treated with CAP at 1 L/min. CAP treatment of SK-OV-3 significantly reduces viability at nearly all doses tested using 3 L/min (Figure 4A) and at all doses using 1 L/min flow rate (Figure 4B). * $p \leq 0.05$.

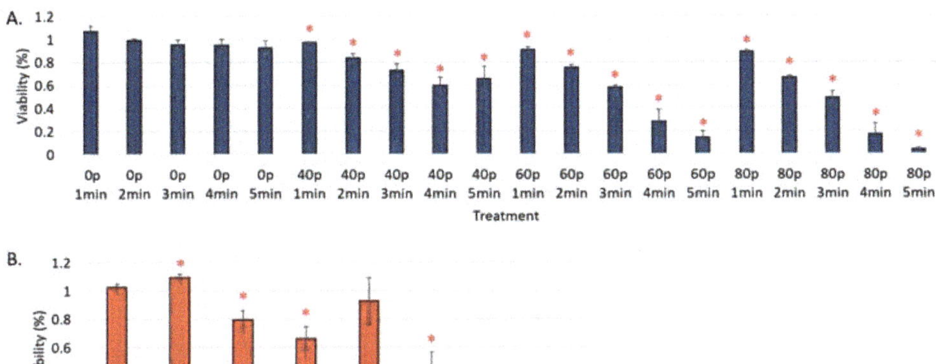

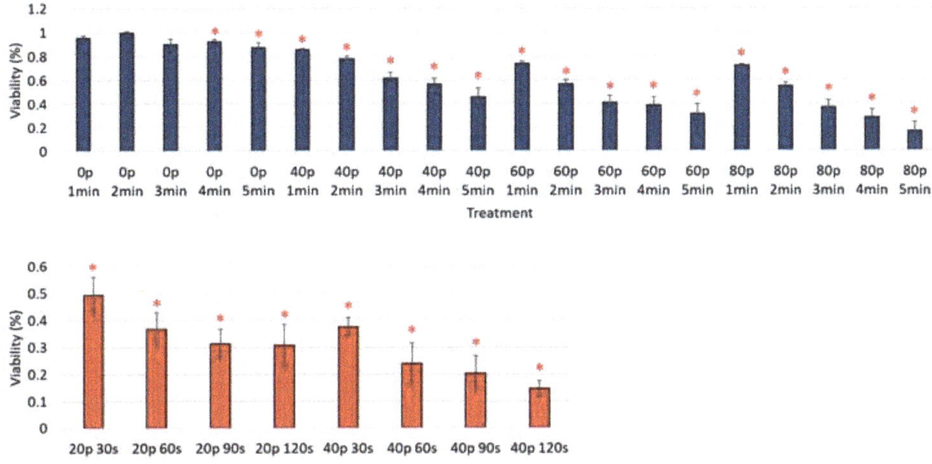

Figure 6. The reduction of the viability of BxPC-3 (pancreatic adenocarcinoma) following CAP treatment. (**A**) BxPC-3 cells were cultured in 12-well plates and treated with CAP at 3 L/min; (**B**) BxPC-3 cells were cultured in 96-well plates and treated with CAP at 1 L/min. CAP treatment of BxPC-3 significantly reduces viability at all doses tested using 3 L/min (Figure 4A) and at nearly all doses using 1 L/min flow rate (Figure 4B). * $p \leq 0.05$.

Figure 7. The reduction of the viability of OE33 (esophageal adenocarcinoma) following CAP treatment. (**top**) OE33 cells were cultured in 12-well plates and treated with CAP at 3 L/min; (**bottom**) OE33 cells were cultured in 96-well plates and treated with CAP at 1 L/min. CAP treatment of OE33 significantly reduces viability at all doses tested, at both 3 L/min and 1 L/min. * $p \leq 0.05$.

4. Discussion

This study is the first effort to describe the dose-dependent reduction of viability, as a combination of treatment time and power settings, on multiple malignant solid tumor cell lines using the USMI Cold Plasma Conversion Unit and Canady Helios Cold Plasma Scalpel. This CAP does not induce any damage on normal tissue, which is described in detail in a separate paper [39]. CAP treatment consistently resulted in a dose-dependent reduction in viability on all solid tumor cell lines tested. While the lowest effective dose varied from cell line to cell line, in each case an 80–99% reduction in viability was achievable 48 h after CAP treatment. 769-P and HCT-116 required a lower dose of plasma while SK-OV-3, BxPC-3, and OE33 required a relatively higher dose. In all cell lines tested, helium treatment alone (0 P) showed no decrease in viability, indicating that the observed effects are due to CAP. While in several of the cell lines 1 L/min flow rate resulted in a lower viability at a lower dose, this cannot be directly compared with the 3 L/min results. This is because the treatment conditions, including well size, beam length, media volume, and cell number, are different between these two assays, and Yan et al. demonstrated that those conditions can alter the effect of CAP on cell viability [17,20]. Xu et al. have suggested a formula to compare plasma dose among treatment conditions within one cell type and this may be necessary to compare future results [40].

Ma et al. demonstrated that the effectiveness of non-thermal plasma treatment was partially dependent on p53 expression [19]. The viability of cancer cells lacking p53 was significantly reduced by non-thermal plasma treatment while p53$^+$ cells were less affected. It is thought that this is due to the role of p53 in protecting the cell from reactive oxygen species [41]. However, based on the established literature, the cell lines tested here, except for SK-OV-3, have been shown to be positive for p53 (Table 1). Ma et al. also used HCT-116 and surprisingly found only a slight reduction in viability whereas we found that viability was reduced to as low as 3.6% (Figure 4A). This is likely due to differences in plasma generation, flow rate, and assay timing. The cell lines tested here also include both wild-type (WT) and mutant p53 (Table 1). Despite positive p53 expression, or status, the viability of these cell types was significantly affected by CAP treatment. However, the cell lines with wild-type p53 (769-P/Figure 3, HCT-116/Figure 4) tended to require a lower dose of CAP to reduce viability compared to those with mutant p53 (BxPC-3/Figure 6, OE33/Figure 7).

Table 1. Status and expression of p53 in all cell types tested; wild-type (WT) and mutant (MUT).

Cell Name	Cell Type	p53 Status	P53 Expression	Reference
769-P	Renal adenocarcinoma	WT	Positive	[42–45]
HCT-116	Colorectal carcinoma	WT	Positive	[19,46]
SK-OV-3	Ovarian adenocarcinoma	MUT/NULL	Negative	[45–48]
BxPC-3	Pancreatic adenocarcinoma	MUT	Positive	[49–51]
OE33	Esophageal adenocarcinoma	MUT	Positive	[45,52]

Further experiments will investigate the effect of CAP on additional cancerous cell lines as well as normal tissues to demonstrate the safety of our cold plasma device. The potential of CAP and chemotherapy combined therapies has been reported repeatedly [53–55]. For cell lines that require a higher dose of CAP to effectively reduce viability, that dose may be reduced with the addition of chemotherapy drugs. Lee et al. demonstrated that CAP can overcome drug resistance in breast cancer [56] and a similar combination therapy may also further reduce the viability of our cell lines. Future experiments will also include preclinical murine models to determine the efficacy of CAP on tumor reduction and prevention of recurrence. Tumors would be treated directly with CAP and tumor size and survival would be measured. To determine the effect of CAP in preventing tumor recurrence; the tumor would be surgically removed and the resulting margins would be treated with CAP. These experiments would match the approach taken in a surgical setting and could lead to improved patient outcomes.

5. Conclusions

This study is the first effort to describe the dose-dependent reduction of viability on multiple malignant solid tumor cell lines using the USMI Cold Plasma Conversion Unit and Canady Helios Cold Plasma Scalpel. The impact of cold plasma on the reduction of viability was consistently dose-dependent and, in each case, an 80–99% reduction in viability was achievable 48 h after CAP treatment. These data demonstrate that the dose required to reduce viability was variable between cell lines; therefore, it is important to select the appropriate CAP dose necessary for treating a specific cancer cell type. This study will provide a cell line-specific dose estimation for future preclinical and clinical studies.

Supplementary Materials: Supplementary Materials are available online at http://www.mdpi.com/2571-6182/1/1/16/s1.

Author Contributions: Conceptualization, W.R., X.C., G.B., B.T., M.K. and J.C.; Data curation, W.R., X.C., L.L. and T.Z.; Formal analysis, W.R., X.C.; Funding acquisition, J.C.; Investigation, W.R., X.C. and L.L.; Methodology, W.R., X.C., M.K. and J.C.; Project administration, W.R. and X.C.; Software, W.R., X.C. and T.Z.; Supervision, M.K. and J.C.; Validation, W.R. and X.C.; Writing—original draft, W.R. and X.C.; Writing—review & editing, W.R., X.C., L.L., T.Z., G.B., B.T., M.K. and J.C.

Acknowledgments: The authors would like to thank the engineering team at Plasma Medicine Life Sciences for technical support of the plasma unit. This research was funded by US Medical Innovations.

Conflicts of Interest: The authors declare no conflict of interest.

References

1. Turhal, N. Two cases of advanced renal cell cancer with prolonged survival of 8 and 12 years. *Jpn. J. Clin. Oncol.* **2002**, *32*, 152–153. [CrossRef] [PubMed]

2. Dabestani, S.; Beisland, C.; Stewart, G.D.; Bensalah, K.; Gudmundsson, E.; Lam, T.B.; Gietzmann, W.; Zakikhani, P.; Marconi, L.; Fernandez-Pello, S.; et al. Long-term outcomes of follow-up for initially localised clear cell renal cell carcinoma: Recur database analysis. *Eur. Urol. Focus* **2018**. [CrossRef] [PubMed]

3. Wille-Jorgensen, P.; Syk, I.; Smedh, K.; Laurberg, S.; Nielsen, D.T.; Petersen, S.H.; Renehan, A.G.; Horvath-Puho, E.; Pahlman, L.; Sorensen, H.T.; et al. Effect of more vs less frequent follow-up testing on overall and colorectal cancer-specific mortality in patients with stage ii or iii colorectal cancer: The colofol randomized clinical trial. *JAMA* **2018**, *319*, 2095–2103. [CrossRef] [PubMed]

4. Benzel, J.; Fendrich, V. Chemoprevention and treatment of pancreatic cancer: Update and review of the literature. *Digestion* **2018**, *97*, 275–287. [CrossRef] [PubMed]

5. Nakayama, Y.; Sugimoto, M.; Gotohda, N.; Konishi, M.; Takahashi, S. Efficacy of completion pancreatectomy for recurrence of adenocarcinoma in the remnant pancreas. *J. Surg. Res.* **2018**, *221*, 15–23. [CrossRef] [PubMed]

6. Conroy, T.; Desseigne, F.; Ychou, M.; Bouche, O.; Guimbaud, R.; Becouarn, Y.; Adenis, A.; Raoul, J.L.; Gourgou-Bourgade, S.; de la Fouchardiere, C.; et al. Folfirinox versus gemcitabine for metastatic pancreatic cancer. *N. Engl. J. Med.* **2011**, *364*, 1817–1825. [CrossRef] [PubMed]

7. Torre, L.A.; Trabert, B.; DeSantis, C.E.; Miller, K.D.; Samimi, G.; Runowicz, C.D.; Gaudet, M.M.; Jemal, A.; Siegel, R.L. Ovarian cancer statistics, 2018. *CA Cancer J. Clin.* **2018**, *68*. [CrossRef] [PubMed]

8. Hou, M.M.; Chen, Y.; Wu, Y.K.; Xi, M.R. Pathological characteristics and prognosis of 664 patients with epithelial ovarian cancer: A retrospective analysis. *J. Sichuan Univ. Med. Sci. Ed.* **2014**, *45*, 859–862.

9. Visser, E.; Edholm, D.; Smithers, B.M.; Thomson, I.G.; Burmeister, B.H.; Walpole, E.T.; Gotley, D.C.; Joubert, W.L.; Atkinson, V.; Mai, T.; et al. Neoadjuvant chemotherapy or chemoradiotherapy for adenocarcinoma of the esophagus. *J. Surg. Oncol.* **2018**. [CrossRef] [PubMed]

10. Xi, M.; Yang, Y.; Zhang, L.; Yang, H.; Merrell, K.W.; Hallemeier, C.L.; Shen, R.K.; Haddock, M.G.; Hofstetter, W.L.; Maru, D.M.; et al. Multi-institutional analysis of recurrence and survival after neoadjuvant chemoradiotherapy of esophageal cancer: Impact of histology on recurrence patterns and outcomes. *Ann. Surg.* **2018**. [CrossRef] [PubMed]

11. Arora, V. Cold Atmospheric Plasma (CAP) in dentistry. *Dentistry* **2013**, *4*. [CrossRef]

12. Niemira, B.A.; Boyd, G.; Sites, J. Cold plasma rapid decontamination of food contact surfaces contaminated with salmonella biofilms. *J. Food Sci.* **2014**, *79*, M917–M922. [CrossRef] [PubMed]

13. Schmidt, A.; Woedtke, T.V.; Stenzel, J.; Lindner, T.; Polei, S.; Vollmar, B.; Bekeschus, S. One year follow-up risk assessment in skh-1 mice and wounds treated with an argon plasma jet. *Int. J. Mol. Sci.* **2017**, *18*, 868. [CrossRef] [PubMed]

14. Shahbazi Rad, Z.; Abbasi Davani, F. Non-thermal atmospheric pressure dielectric barrier discharge plasma source construction and investigation on the effect of grid on wound healing application. *Clin. Plasma Med.* **2016**, *4*, 56–64. [CrossRef]

15. Volotskova, O.; Dubrovsky, L.; Keidar, M.; Bukrinsky, M. Cold atmospheric plasma inhibits hiv-1 replication in macrophages by targeting both the virus and the cells. *PLoS ONE* **2016**, *11*, e0165322. [CrossRef] [PubMed]

16. Keidar, M. Plasma for cancer treatment. *Plasma Sources Sci. Technol.* **2015**, *24*, 033001. [CrossRef]

17. Yan, D.; Talbot, A.; Nourmohammadi, N.; Cheng, X.; Canady, J.; Sherman, J.; Keidar, M. Principles of using cold atmospheric plasma stimulated media for cancer treatment. *Sci. Rep.* **2015**, *5*, 18339. [CrossRef] [PubMed]

18. Naciri, M.; Dowling, D.; Al-Rubeai, M. Differential sensitivity of mammalian cell lines to non-thermal atmospheric plasma. *Plasma Process. Polym.* **2014**, *11*, 391–400. [CrossRef]

19. Ma, Y.; Ha, C.S.; Hwang, S.W.; Lee, H.J.; Kim, G.C.; Lee, K.W.; Song, K. Non-thermal atmospheric pressure plasma preferentially induces apoptosis in p53-mutated cancer cells by activating ros stress-response pathways. *PLoS ONE* **2014**, *9*, e91947. [CrossRef] [PubMed]

20. Yan, D.; Sherman, J.H.; Canady, J.; Trink, B.; Keidar, M. The cellular ros-scavenging function, a key factor determining the specific vulnerability of cancer cells to cold atmospheric plasma in vitro. *arXiv* **2017**, arXiv:1711.09015.

21. Lee, H.J.; Shon, C.H.; Kim, Y.S.; Kim, S.; Kim, G.C.; Kong, M.G. Degradation of adhesion molecules of g361 melanoma cells by a non-thermal atmospheric pressure microplasma. *New J. Phys.* **2009**, *11*, 115026. [CrossRef]

22. Schmidt, A.; Bekeschus, S.; von Woedtke, T.; Hasse, S. Cell migration and adhesion of a human melanoma cell line is decreased by cold plasma treatment. *Clin. Plasma Med.* **2015**, *3*, 24–31. [CrossRef]

23. Shi, X.-M.; Zhang, G.-J.; Chang, Z.-S.; Wu, X.-L.; Liao, W.-L.; Li, N. Viability reduction of melanoma cells by plasma jet via inducing g1/s and g2/m cell cycle arrest and cell apoptosis. *IEEE Trans. Plasma Sci.* **2014**, *42*, 1640–1647. [CrossRef]

24. Chang, J.W.; Kang, S.U.; Shin, Y.S.; Kim, K.I.; Seo, S.J.; Yang, S.S.; Lee, J.S.; Moon, E.; Baek, S.J.; Lee, K.; et al. Non-thermal atmospheric pressure plasma induces apoptosis in oral cavity squamous cell carcinoma: Involvement of DNA-damage-triggering sub-g(1) arrest via the atm/p53 pathway. *Arch. Biochem. Biophys.* **2014**, *545*, 133–140. [CrossRef] [PubMed]

25. Gherardi, M.; Turrini, E.; Laurita, R.; De Gianni, E.; Ferruzzi, L.; Liguori, A.; Stancampiano, A.; Colombo, V.; Fimognari, C. Atmospheric non-equilibrium plasma promotes cell death and cell-cycle arrest in a lymphoma cell line. *Plasma Process. Polym.* **2015**, *12*, 1354–1363. [CrossRef]

26. Adachi, T.; Tanaka, H.; Nonomura, S.; Hara, H.; Kondo, S.; Hori, M. Plasma-activated medium induces a549 cell injury via a spiral apoptotic cascade involving the mitochondrial-nuclear network. *Free Radic. Biol. Med.* **2015**, *79*, 28–44. [CrossRef] [PubMed]

27. Ahn, H.J.; Kim, K.I.; Kim, G.; Moon, E.; Yang, S.S.; Lee, J.S. Atmospheric-pressure plasma jet induces apoptosis involving mitochondria via generation of free radicals. *PLoS ONE* **2011**, *6*, e28154. [CrossRef] [PubMed]

28. Kim, S.J.; Chung, T.H.; Bae, S.H.; Leem, S.H. Induction of apoptosis in human breast cancer cells by a pulsed atmospheric pressure plasma jet. *Appl. Phys. Lett.* **2010**, *97*, 023702. [CrossRef]

29. Keidar, M.; Walk, R.; Shashurin, A.; Srinivasan, P.; Sandler, A.; Dasgupta, S.; Ravi, R.; Guerrero-Preston, R.; Trink, B. Cold plasma selectivity and the possibility of a paradigm shift in cancer therapy. *Br. J. Cancer* **2011**, *105*, 1295–1301. [CrossRef] [PubMed]

30. Nuccitelli, R.; Chen, X.; Pakhomov, A.G.; Baldwin, W.H.; Sheikh, S.; Pomicter, J.L.; Ren, W.; Osgood, C.; Swanson, R.J.; Kolb, J.F.; et al. A new pulsed electric field therapy for melanoma disrupts the tumor's blood supply and causes complete remission without recurrence. *Int. J. Cancer* **2009**, *125*, 438–445. [CrossRef] [PubMed]

31. Chung, W.H. Mechanisms of a novel anticancer therapeutic strategy involving atmospheric pressure plasma-mediated apoptosis and DNA strand break formation. *Arch. Pharm. Res.* **2016**, *39*, 1–9. [CrossRef] [PubMed]

32. Canady, J.; Shashurin, A.; Keidar, M.; Zhuang, T. Integrated Cold Plasma and High Frequency Plasma Electrosurgical System and Method. U.S. Patent No. 9,999,462, 19 June 2018.

33. Okazaki, S.; Kogoma, M.; Uehara, M.; Kimura, Y. Appearance of stable glow discharge in air, argon, oxygen and nitrogen at atmospheric pressure using a 50 hz source. *J. Phys. D* **1993**, *26*, 889–892. [CrossRef]

34. Sublet, A.; Ding, C.; Dorier, J.L.; Hollenstein, C.; Fayet, P.; Coursimault, F. Atmospheric and sub-atmospheric dielectric barrier discharges in helium and nitrogen. *Plasma Sources Sci. Technol.* **2006**, *15*, 627–634. [CrossRef]

35. Cheng, X.; Rowe, W.J.; Ly, L.; Shashurin, A.; Zhuang, T.; Wigh, S.; Basadonna, G.; Trink, B.; Keidar, M.; Canady, J. Treatment of triple-negative breast cancer cells with the canady cold plasma conversion system: Preliminary results. *Plasma* **2018**, submitted.

36. Pearse, R.W.B.; Gaydon, A.G. *The Identification of Molecular Spectra*; Chapman and Hall: London, UK, 1976.

37. Liu, Z.; Xu, D.; Liu, D.; Cui, Q.; Cai, H.; Li, Q.; Chen, H.; Kong, M.G. Production of simplex rns and ros by nanosecond pulse N_2/O_2 plasma jets with homogeneous shielding gas for inducing myeloma cell apoptosis. *J. Phys. D Appl. Phys.* **2017**, *50*, 195204. [CrossRef]

38. Lunov, O.; Zablotskii, V.; Churpita, O.; Lunova, M.; Jirsa, M.; Dejneka, A.; Kubinova, S. Chemically different non-thermal plasmas target distinct cell death pathways. *Sci. Rep.* **2017**, *7*, 600. [CrossRef] [PubMed]

39. Ly, L.; Jones, S.; Shashurin, A.; Zhuang, T.; Rowe, W.; Cheng, X.; Wigh, S.; Naab, T.; Keidar, M.; Canady, J. A new cold plasma jet: Performance evaluation of cold plasma, hybrid plasma and argon plasma coagulation. *Plasma* **2018**, submitted.

40. Xu, X.; Dai, X.; Xiang, L.; Cai, D.; Xiao, S.; Ostrikov, K. Quantitative assessment of cold atmospheric plasma anti-cancer efficacy in triple-negative breast cancers. *Plasma Process. Polym.* **2018**, *15*. [CrossRef]

41. Sablina, A.A.; Budanov, A.V.; Ilyinskaya, G.V.; Agapova, L.S.; Kravchenko, J.E.; Chumakov, P.M. The antioxidant function of the p53 tumor suppressor. *Nat. Med.* **2005**, *11*, 1306–1313. [CrossRef] [PubMed]

42. Wang, J.; Zhang, P.; Zhong, J.; Tan, M.; Ge, J.; Tao, L.; Li, Y.; Zhu, Y.; Wu, L.; Qiu, J.; et al. The platelet isoform of phosphofructokinase contributes to metabolic reprogramming and maintains cell proliferation in clear cell renal cell carcinoma. *Oncotarget* **2016**, *7*, 27142–27157. [CrossRef] [PubMed]

43. Miyazaki, J.; Ito, K.; Fujita, T.; Matsuzaki, Y.; Asano, T.; Hayakawa, M.; Asano, T.; Kawakami, Y. Progression of human renal cell carcinoma via inhibition of rhoa-rock axis by PARG1. *Transl. Oncol.* **2017**, *10*, 142–152. [CrossRef] [PubMed]

44. Mu, W.; Hu, C.; Zhang, H.; Qu, Z.; Cen, J.; Qiu, Z.; Li, C.; Ren, H.; Li, Y.; He, X.; et al. Mir-27b synergizes with anticancer drugs via p53 activation and CYP1B1 suppression. *Cell Res.* **2015**, *25*, 477–495. [CrossRef] [PubMed]

45. Bamford, S.; Dawson, E.; Forbes, S.; Clements, J.; Pettett, R.; Dogan, A.; Flanagan, A.; Teague, J.; Futreal, P.A.; Stratton, M.R.; et al. The cosmic (catalogue of somatic mutations in cancer) database and website. *Br. J. Cancer* **2004**, *91*, 355–358. [CrossRef] [PubMed]

46. O'Connor, P.M.; Jackman, J.; Bae, I.; Myers, T.G.; Fan, S.; Mutoh, M.; Scudiero, D.A.; Monks, A.; Sausville, E.A.; Weinstein, J.N.; et al. Characterization of the p53 tumor suppressor pathway in cell lines of the national cancer institute anticancer drug screen and correlations with the growth-inhibitory potency of 123 anticancer agents. *Cancer Res.* **1997**, *57*, 4285–4300. [PubMed]

47. Antoun, S.; Atallah, D.; Tahtouh, R.; Alaaeddine, N.; Moubarak, M.; Khaddage, A.; Ayoub, E.N.; Chahine, G.; Hilal, G. Different tp53 mutants in p53 overexpressed epithelial ovarian carcinoma can be associated both with altered and unaltered glycolytic and apoptotic profiles. *Cancer Cell Int.* **2018**, *18*, 14. [CrossRef] [PubMed]

48. Yaginuma, Y.; Westphal, H. Abnormal structure and expression of the p53 gene in human ovarian carcinoma cell lines. *Cancer Res.* **1992**, *52*, 4196–4199. [PubMed]

49. Chen, D.; Niu, M.; Jiao, X.; Zhang, K.; Liang, J.; Zhang, D. Inhibition of AKT2 enhances sensitivity to gemcitabine via regulating puma and NF-κB signaling pathway in human pancreatic ductal adenocarcinoma. *Int. J. Mol. Sci.* **2012**, *13*, 1186–1208. [CrossRef] [PubMed]

50. Wang, F.; Li, H.; Yan, X.G.; Zhou, Z.W.; Yi, Z.G.; He, Z.X.; Pan, S.T.; Yang, Y.X.; Wang, Z.Z.; Zhang, X. Alisertib induces cell cycle arrest and autophagy and suppresses epithelial-to-mesenchymal transition involving PI3K/AKT/mTOR and sirtuin 1-mediated signaling pathways in human pancreatic cancer cells. *Drug Des. Dev. Ther.* **2015**, *9*, 575–601.

51. Ruggeri, B.; Zhang, S.Y.; Caamano, J.; DiRado, M.; Flynn, S.D.; Klein-Szanto, A.J. Human pancreatic carcinomas and cell lines reveal frequent and multiple alterations in the p53 and Rb-1 tumor-suppressor genes. *Oncogene* **1992**, *7*, 1503–1511. [PubMed]

52. Liu, D.S.; Read, M.; Cullinane, C.; Azar, W.J.; Fennell, C.M.; Montgomery, K.G.; Haupt, S.; Haupt, Y.; Wiman, K.G.; Duong, C.P.; et al. APR-246 potently inhibits tumour growth and overcomes chemoresistance in preclinical models of oesophageal adenocarcinoma. *Gut* **2015**, *64*, 1506–1516. [CrossRef] [PubMed]

53. Brulle, L.; Vandamme, M.; Ries, D.; Martel, E.; Robert, E.; Lerondel, S.; Trichet, V.; Richard, S.; Pouvesle, J.M.; Le Pape, A. Effects of a non thermal plasma treatment alone or in combination with gemcitabine in a MIA PaCa2-luc orthotopic pancreatic carcinoma model. *PLoS ONE* **2012**, *7*, e52653. [CrossRef] [PubMed]

54. Chen, C.-Y.; Cheng, Y.-C.; Cheng, Y.-J. Synergistic effects of plasma-activated medium and chemotherapeutic drugs in cancer treatment. *J. Phys. D* **2018**, *51*, 13LT01. [CrossRef]

55. Masur, K.; von Behr, M.; Bekeschus, S.; Weltmann, K.-D.; Hackbarth, C.; Heidecke, C.-D.; von Bernstorff, W.; von Woedtke, T.; Partecke, L.I. Synergistic inhibition of tumor cell proliferation by cold plasma and gemcitabine. *Plasma Process. Polym.* **2015**, *12*, 1377–1382. [CrossRef]

56. Lee, S.; Lee, H.; Jeong, D.; Ham, J.; Park, S.; Choi, E.H.; Kim, S.J. Cold atmospheric plasma restores tamoxifen sensitivity in resistant MCF-7 breast cancer cell. *Free Radic. Biol. Med.* **2017**, *110*, 280–290. [CrossRef] [PubMed]

 *plasma*

Article

A New Cold Plasma Jet: Performance Evaluation of Cold Plasma, Hybrid Plasma and Argon Plasma Coagulation

Lawan Ly [1], Sterlyn Jones [1], Alexey Shashurin [1,2], Taisen Zhuang [3], Warren Rowe III [1], Xiaoqian Cheng [1], Shruti Wigh [3], Tammey Naab [4], Michael Keidar [1,5] and Jerome Canady [1,5,*]

[1] Jerome Canady Research Institute for Advanced Biological and Technological Sciences, Takoma Park, MD 20912, USA; llawan@usmedinnov.com (L.L.); sterjones78@aol.com (S.J.); ashashur@purdue.edu (A.S.); drwrowe@usmedinnov.com (W.R.); xcheng@usmedinnov.com (X.C.); keidar@email.gwu.edu (M.K.)

[2] Department of Aerospace Engineering, Purdue University, West Lafayette, IN 47907, USA

[3] Plasma Medicine Life Sciences, Takoma Park, MD 20912, USA; tzhuang@usmedinnov.com (T.Z.); swigh@usmedinnov.com (S.W.)

[4] Department of Pathology, Howard University Hospital, Washington, DC 20059, USA; tjnaab@gmail.com

[5] Department of Mechanical and Aerospace Engineering, The George Washington University, Washington, DC 20052, USA

[*] Correspondence: drjcanady@usmedinnov.com; Tel.: +1-301-270-0147

Received: 2 August 2018; Accepted: 7 September 2018; Published: 11 September 2018

Abstract: The use of plasma energy has expanded in surgery and medicine. Tumor resection in surgery and endoscopy has incorporated the use of a plasma scalpel or catheter for over four decades. A new plasma energy has expanded the tools in surgery: Cold Atmospheric Plasma (CAP). A cold plasma generator and handpiece are required to deliver the CAP energy. The authors evaluated a new Cold Plasma Jet System. The Cold Plasma Jet System consists of a USMI Cold Plasma Conversion Unit, Canady Helios Cold Plasma® Scalpel, and the Canady Plasma® Scalpel in Hybrid and Argon Plasma Coagulation (APC) modes. This plasma surgical system is designed to remove the target tumor with minimal blood loss and subsequently spray the local area with cold plasma. In this study, various operational parameters of the Canady Plasma® Scalpels were tested on ex vivo normal porcine liver tissue. These conditions included various gas flow rates (1.0, 3.0, 5.0 L/min), powers (20, 40, 60 P), and treatment durations (30, 60, 90, 120 s) with argon and helium gases. Plasma length, tissue temperature changes, and depth and eschar injury magnitude measurements resulting from treatment were taken into consideration in the comparison of the scalpels. The authors report that a new cold plasma jet technology does not produce any thermal damage to normal tissue.

Keywords: cold atmospheric plasma; cold plasma device; Hybrid plasma; argon plasma coagulation

1. Introduction

Electrosurgery is the use of high frequency radiofrequency (RF) alternating current (AC) for cutting and coagulation of tissue. Investigators in Europe and the United States started to explore the effect of AC on tissue in the late nineteenth century. In 1893, a French inventor and physicist, Arsene d'Arsonval was the first to report the clinical effect of AC on tissue [1]. Many Europeans and Americans contributed to the development of the AC electrosurgical generator but the invention by Dr. William T. Bovie [2] stands out the most. Bovie's contributions in electrosurgery enhanced after collaborating with Dr. Harvey Cushing, Surgeon in Chief at Peter Bent Brigham, who used Bovie's electrosurgical generator for a brain tumor because of excessive bleeding [3].

Plasma technology in surgery has advanced since the first introduction by Morrison 44 years ago [4]. In 1990, Canady [5] was the first to describe the delivery of argon plasma coagulation via a flexible endoscope. These new methods allowed the surgeon and endoscopist to combine standard monopolar electrocautery with a plasma gas for the coagulation of tissue. In 2010, Canady et al. [6] developed a new mode of plasma- "Hybrid Plasma". Hybrid plasma combines monopolar electrical charge with a plasma gas which creates a plasma beam that can simultaneously cut and coagulate biological tissue.

After the introduction of Canady's Hybrid plasma, we continued to develop several electrosurgical systems. In our previous studies, we tested an older generation of our Hybrid electrosurgical system (SS-200E/Argon 2) using an argon flow rate of 3 L/min and powers of 40, 60, and 80 W [7,8]. Pure cutting [7] and pure coagulation [8] was successfully demonstrated on various animal biological samples. In another study, we compared our SS-200E/Argon 2 and SS-501MCa/Argon 4 electrosurgical systems in terms of the plasma properties and injury characteristics produced [9]. Conventional coagulation, argon coagulation, conventional cut, and argon cut modes were explored with treatment times totaling 5 s with various flow rates and powers. Both electrosurgical systems operating under argon cut mode and gentle parameters exhibited injury sizes suitable for surgical procedures.

We have recently developed a series of electrosurgical systems compromised of the Canady Helios Plasma® Scalpel with Cold Atmospheric Plasma (CAP) mode (operating with the SS-601MCa generator and USMI Cold Plasma Conversion Unit) [10] and the Canady Plasma® Scalpel, which is capable of Hybrid and Argon Plasma Coagulation (APC) modes (operating on the USMI SS-601MCa/Argon 4 generator). The Canady Plasma® Scalpel in Hybrid mode, capable of simultaneous cut and coagulation, has the potential to surgically remove the target tumor while coagulating the surrounding tissue. In addition, the Canady Plasma® Scalpel in APC mode can aid in the procedure by providing pure coagulation. The coagulation capabilities of Hybrid and APC modes could minimize blood loss during the surgical procedure by sealing off blood vessels, which in turn may reduce post-operative complications that arise from hemorrhaging. After the tumor is removed using the Canady Plasma® Scalpel in Hybrid and APC modes, the Canady Helios Plasma® Scalpel can be used to spray the surgical margins with CAP. The treatment with CAP will eliminate undetected cancer cells, thus preventing tumor recurrence.

Recently, CAP has been gaining more interest as a potential cancer treatment due to promising research results. CAP is known to selectively ablate cancer cells in vitro and reduce tumor size in vivo [11,12]. The reactive oxygen and nitrogen species (RONS) generated by CAP has been recognized as a possible factor in selective eradication [12–14]. It has been observed that RONS induce apoptosis in cancer cells through an oxidative DNA damage cascade [15] and downregulation of oxidative stress-related genes [16]. To this end, selective treatment of surgical margins allowing elimination of cancerous areas has tremendous advantage; this could be provided by the Canady Helios Plasma® Scalpel.

In our recent study, we have shown that the spectrum of CAP, generated by the USMI Cold Plasma Conversion Unit and Canady Helios Cold Plasma Scalpel, mainly consists of N_2, N_2^+, He, and O [17]. We have also demonstrated that the CAP jet significantly reduced cell viability in a dose-dependent manner in various cell lines such as 769-P (renal adenocarcinoma), HCT-116 (colorectal carcinoma), SK-OV-3 (ovarian adenocarcinoma), BxPC-3 (pancreatic adenocarcinoma), OE33 (oesophageal adenocarcinoma), and MDA-MB-231 (breast adenocarcinoma) [17,18]. With various CAP dosages, an 80–99% reduction in viability of these cancer cells lines was achieved 48 h after treatment [17]. However, the safety of this new device must be taken into consideration.

In this study, we evaluated the extent of the thermal injuries produced by the Canady Plasma® Scalpels in combination with our new generation of generators (SS-601MCa generator/USMI Cold Plasma Conversion Unit and USMI SS-601MCa/Argon 4 generator) (US Medical Innovations, Takoma Park, MD, USA). Various parameters of the Canady Plasma® Scalpels were tested to compare the severity of injuries produced by CAP, Hybrid, and APC modes.

2. Materials and Methods

2.1. Canady Plasma® Scalpels

All experiments were performed at the Jerome Canady Research Institute for Advanced Biological and Technological Sciences, (JCRI-ABTS), Takoma Park, MD, USA. All experiments were conducted using the Canady Helios Cold Plasma® Scalpel and the Canady Plasma Scalpel in Hybrid and APC Mode, SS-601/Cold Plasma Converter (US Medical Innovations, LLC., Takoma Park, MD, USA), pictured in Figure 1. Each Canady Plasma® Scalpel has a channel and an electrode with an opening at the end. An inert gas flows through the channel to the electrode where a high-frequency energy is applied to the electrode causing the gas to ionize as it exits the port. Helium gas is used for the Canady Helios Plasma® Scalpel and argon gas for the Canady Plasma® Scalpel in Hybrid and APC modes.

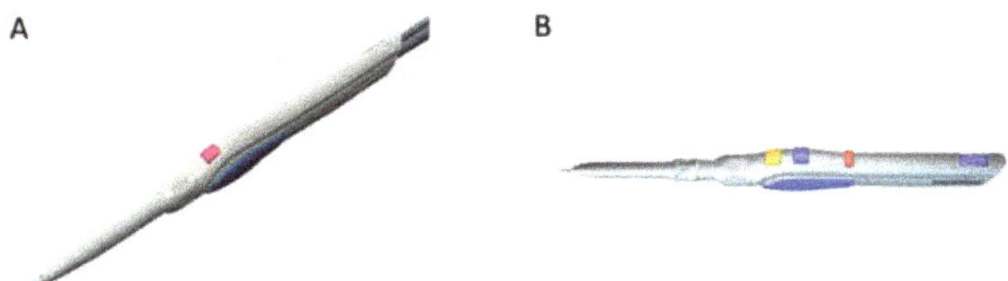

Figure 1. (**A**) Canady Helios Cold Plasma® Scalpel and (**B**) Canady Plasma® Scalpel with Hybrid and Argon Plasma Coagulation (APC) Modes.

2.1.1. CAP Mode

The Canady Helios Plasma® Scalpel utilizes the USMI SS-601MCa generator with a USMI Cold Plasma Conversion Unit to produce CAP. The conversion unit up-converts voltage (1.5–50 kV), down-converts frequency (<300 kHz), and down-converts power (<30 W).

2.1.2. APC Mode

The Candy Plasma® Scalpel in APC mode operates on the Hybrid electrosurgical system which includes the USMI SS-601MCa/Argon 4. For pure coagulation, only the Argon 4 generator is required. Coagulation requires a high voltage waveform with a low duty cycle. The surges in voltage power, referred to as spikes, are necessary for coagulation. When the surrounding tissue is heated during the spike and cooled between spikes, tissue coagulation is achieved.

2.1.3. Hybrid Mode

Hybrid mode relies on the USMI SS-601MCa/Argon 4 for simultaneous cutting and coagulation. The Argon 4 coagulator provides the argon gas. To first initiate a cutting action, a sufficient amount of power is required for the formation of a steam layer between the electrode and the tissue. The steam layer subsequently allows for the formation of a plasma composed of highly ionized argon and water. A radiofrequency arc, characterized by high power density, is then developed in the plasma. Upon contact with the tissue, the arc causes immediate disruption in tissue structure. A low voltage waveform with high duty cycle is required to maintain a constant succession of arcs to produce a cut in the tissue.

To achieve simultaneous cutting and coagulation under Hybrid mode, the continuous sinusoidal voltage used for cutting is periodically interrupted. During interruptions, the ionized argon particles in the plasma disperse allowing the electrode to temporarily make direct contact with the tissue. This interaction causes coagulation.

2.2. Preparation and Treatment Conditions of Liver Tissue Samples

All experiments were conducted on ex vivo normal porcine liver tissue. Tissue was divided into 1.5 cm × 1.5 cm samples and kept in a 15 °C environment prior to treatment. Samples were treated with gas flow rates of 1.0, 3.0, and 5.0 L/min. Power settings were tested at 20, 40, and 60 P which corresponded to 5, 8, 11 W when using the USMI Cold Plasma® Conversion Unit and 20, 40, 60 W on the Argon 4/SS-601MCa. Exposure times to the plasma were set to 30, 60, 90, and 120 s. Equipment set up is displayed in Figure 2.

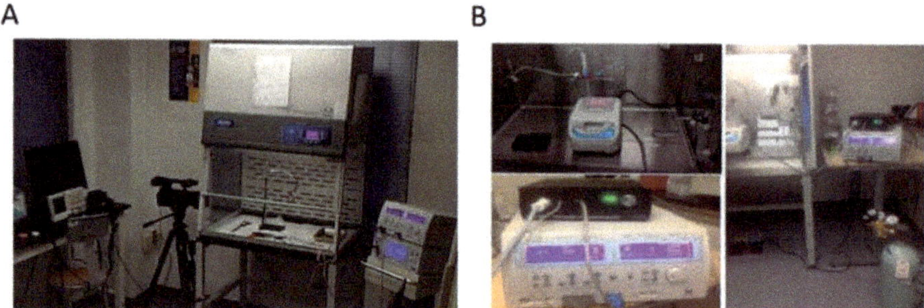

Figure 2. Overall equipment setup with the USMI SS-601MCa generator/USMI Cold Plasma Conversion Unit in (**A**) and the USMI SS-601MCa/USMI Argon 4 generator in (**B**).

Plasma lengths of the Canady Plasma® Scalpels were measured and defined as the maximum length of the plasma beam at which the discharge was sustained. The treatments were video recorded by digital camera (Panasonic AG, Panasonic, Kadoma, Osaka Prefecture, Japan).

Tissue temperature was recorded with infrared temperature sensor FLIR (E Series) before and after treatment. The accuracy of the temperature measurements was less than 12 °C.

The Escher diameter of the injury was measured with a 6-inch digital caliper post-treatment. The accuracy of the diameter measurements was less than 0.05 mm.

The depth of injury was determined by examining stained slides of treated tissue samples. Preparation of the slides began immediately after plasma treatment; samples were fixed in 4% buffered formalin for 5 h, paraffin embedded, and stained for hematoxylin and eosin. Preparation of the pathological slides was performed at the Histopathology Laboratories: Howard University Hospital, Washington, DC, USA and JCRI-ABTS, Takoma Park, MD, USA. The slides were studied under the Zeiss Primovert phase contrast microscope with Zeiss ZEN SP1/SP2 blue edition software (Zeiss, Oberkochen, Germany). Data for plasma length, tissue temperature change, eschar width, and depth of injury for all treatment conditions has been included as Supplementary Materials.

2.3. Statistics

All treatment conditions were repeated 3 times with the exception for plasma length on Hybrid mode (which were negligible and repeated once) and for depth of injury on Hybrid and APC modes (which were repeated 1–3 times). A non-parametric student two-tailed, unpaired t test was performed for all samples $n = 3$. * p-value ≤ 0.05 was considered significant and a mean difference at 95% confidence level was used with SPSS software (IBM, New York, NY, USA).

3. Results

3.1. Plasma Length Measurements

Figure 3 displays plasma discharge produced by each mode. Hybrid plasma is initiated upon contact as seen in Figure 3A. And unlike Hybrid mode, CAP (Figure 3B) and APC (Figure 3C) modes can be held a distance away from the tissue thus do not require contact for plasma discharge.

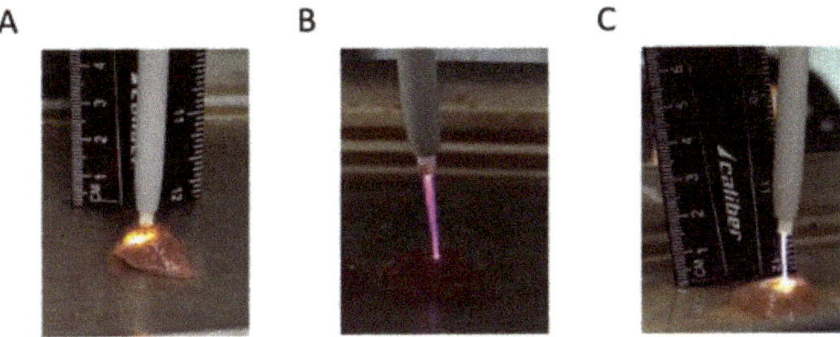

Figure 3. Plasma discharge in (**A**) Hybrid; (**B**) CAP; and (**C**) APC modes.

Figure 4 focuses on the solely on the CAP beam. The discharge elongates with flow rate as the beam lengths increase in ascending order of 1 L/min (Figure 4A), 3 L/min (Figure 4B), and 5 L/min (Figure 4C). Light intensity emitted from the discharge strengthens with power, shown from left to right in all rows of Figure 4.

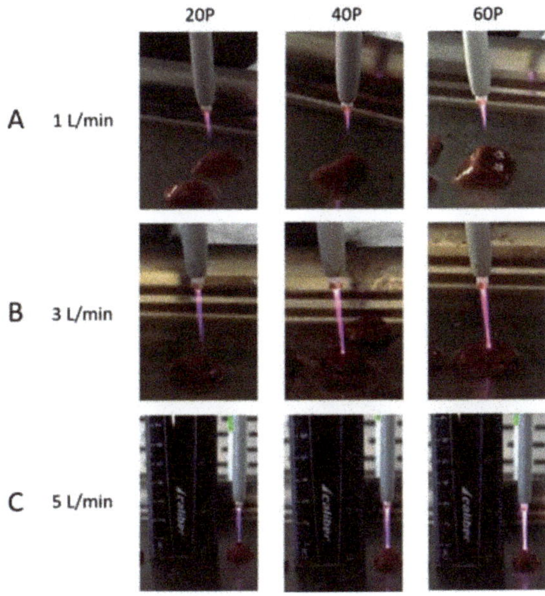

Figure 4. All treatments with the Canady Helios Cold Plasma® Scalpel were maintained at a 2-cm distance from the tissue sample. The CAP beam transformed in length and light intensity under 1 L/min (**A**); 3 L/min (**B**); and 5 L/min (**C**) flow rates and 20, 40, 60 P.

The maximum plasma length generated by each electrosurgical scalpel using various gas flow rates, powers, and treatment durations is shown in Figure 5. The Canady Helios Cold Plasma® Scalpel was constantly held 2 cm away from the tissue surface, hence, the plasma length for cold plasma mode was consistently recorded at 2 cm. Plasma length for Hybrid mode was negligible at 0.025 to 0.45 mm for all conditions. Argon mode demonstrated the most flexibility in plasma length which ranged from 5–20 mm. In the case of APC, one can see that the plasma column length varies with power applied (Figure 5A–I).

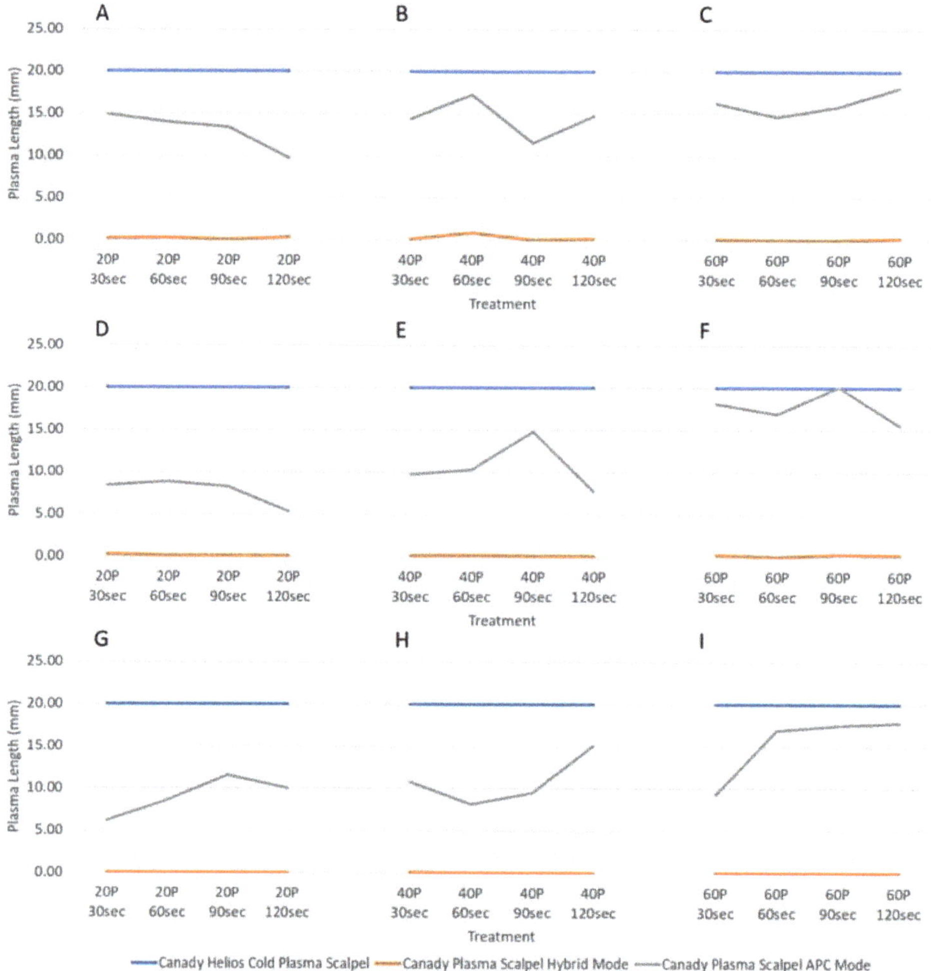

Figure 5. Plasma length of each scalpel was measured during treatments. Treatment conditions were 1 L/min (**A–C**), 3 L/min (**D–F**), and 5 L/min (**G–I**) flow rates with different powers and durations.

3.2. Temperature Change Measurements

Initial temperatures for all tissue samples were estimated to be 15 °C. Total temperature changes in result of the three modes of plasma treatments are displayed in Figure 6. It was clearly demonstrated that both CAP and Hybrid modes produced significantly lower tissue temperatures than APC modes despite increasing plasma dosages (Figure 6A–F). Argon mode proved to induce the greatest thermal effect as final temperatures ascended to 8 to 10-fold at 5 L/min (Figure 6G–I). Under 1 and 3 L/min, it is difficult to establish a definite relationship between the operational parameters of APC and temperature change (Figure 6A–F). However, it was observed that 5 L/min yielded higher final tissue temperatures with longer treatment times until a plateau was reached (Figure 6G–I).

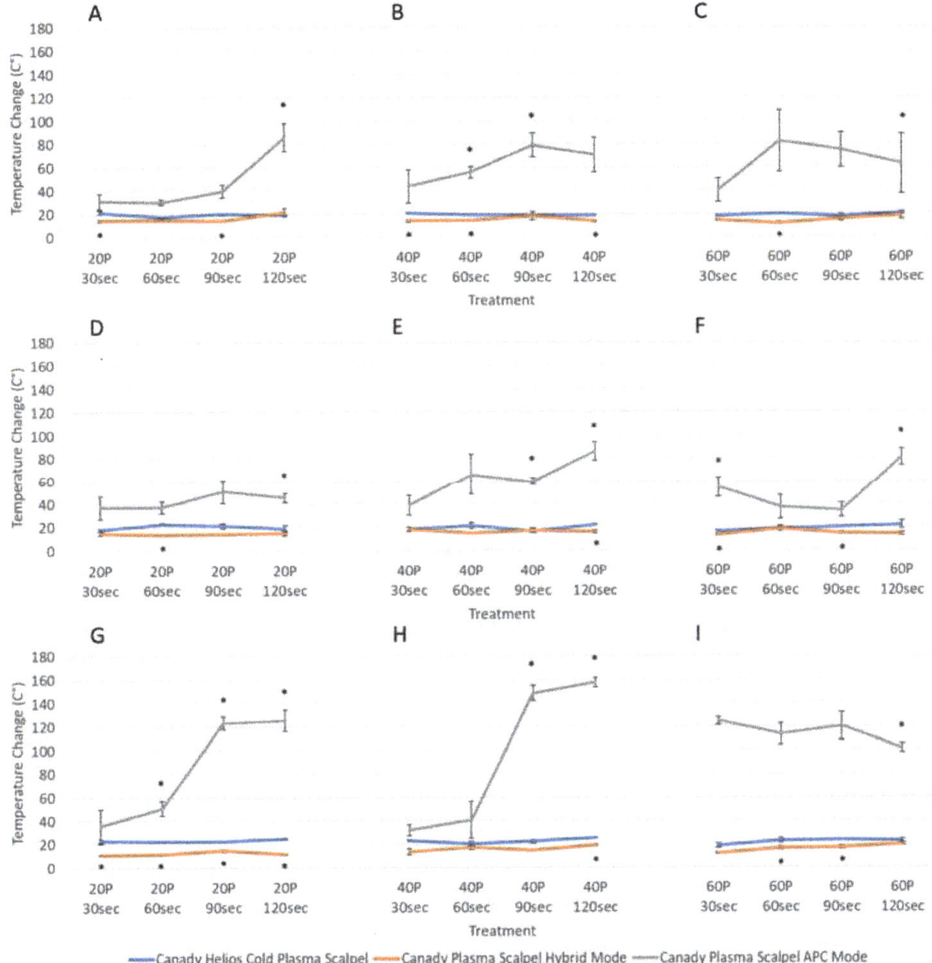

Figure 6. Total temperature changes produced by each scalpel was measured post-treatment. Treatment conditions were 1 L/min (**A–C**), 3 L/min (**D–F**), and 5 L/min (**G–I**) flow rates with different powers and durations.

3.3. Variability in Injury Magnitude: Eschar Diameter

Post-treatment tissue morphology in result of Hybrid mode and APC modes using 3.0 L/min and 40 P for 90 s are shown Figure 7. It is demonstrated that even under the same condition, injury magnitude strongly differs between the two modes. One can observe that the injury produced by Hybrid mode (Figure 7A) is both smaller in depth and eschar diameter and lighter in color than that produced by APC mode (Figure 7B), due to the vast difference in plasma discharge temperature. It can be assumed that dramatic temperature heating leads to greater cell death which is reflected in the greater injury magnitude and darkening of tissue color produced by APC.

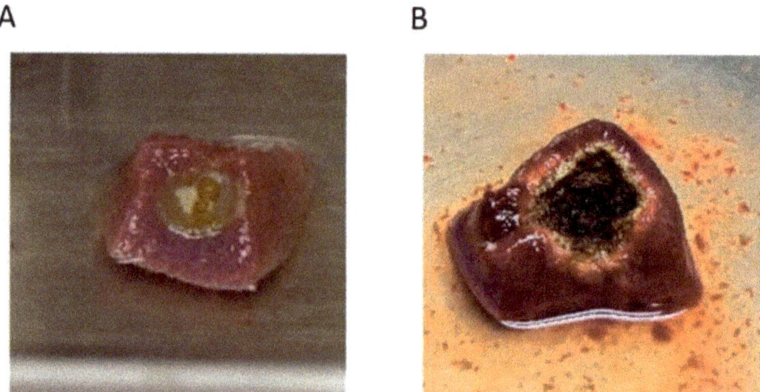

Figure 7. Demonstration of injury magnitude by Hybrid mode (**A**) and APC (**B**) mode using 3.0 L/min and 40 P for a total of 90 s.

Figure 8 presents samples of the pathology images taken after CAP treatment. It shows no eschar diameter injury produced by CAP treatment with settings of 1 (Figure 8A), 3 (Figure 8B), 5 (Figure 8C) L/min at 40 P and 60 s.

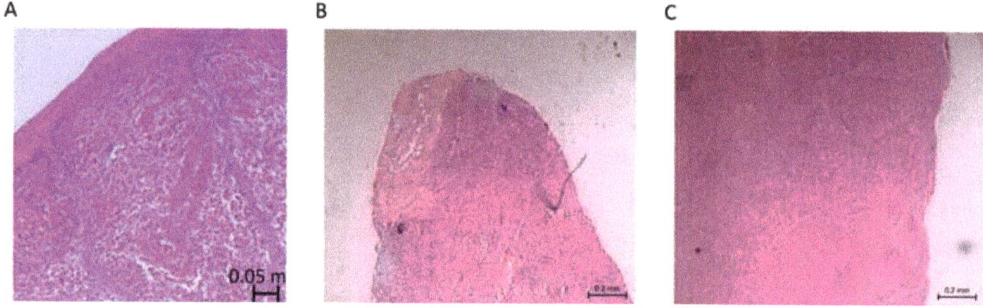

Figure 8. The CAP treatment with 1 L/min (**A**); 3 L/min (**B**); 5 L/min (**C**) and 40 P for 60 s causes no eschar diameter and depth of injury.

Figure 9 presents the average eschar diameter induced by each electrosurgical device. CAP showed no thermal injury as the diameter of injury remained at zero regardless of treatment condition. Injury by Hybrid and APC mode ranged from 1–8 mm and 8–18 mm in diameter, respectively. In both Hybrid and APC mode, change in diameter of injury appears to be independent of gas flow rate. Instead, greater injury is observed with an increase in power. While longer treatment durations with Hybrid plasma seems to have no effect on injury diameter, the results reveal that longer exposure to argon plasma is a factor in the increase of injury diameter.

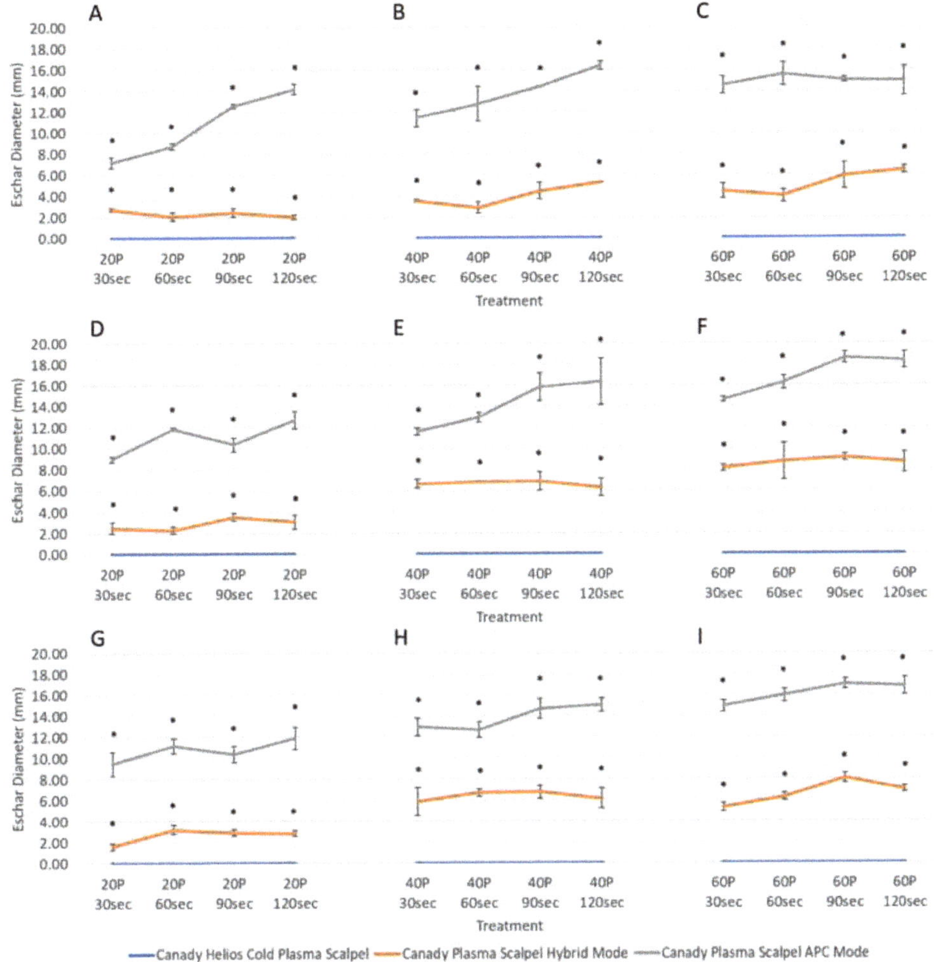

Figure 9. Eschar diameter caused by each scalpel was measured post-treatment. Treatment conditions were 1 L/min (**A–C**), 3 L/min (**D–F**), and 5 L/min (**G–I**) flow rates with different powers and durations.

3.4. Variability in Injury Magnitude: Depth of Injury

Depth of injury under various device parameters are presented in Figure 10. With CAP treatment, depth of injury was 0.0 mm for all conditions (Figures 8A–C and 10A–I). The plasmas generated by Hybrid and APC modes led to relatively greater depth of injury at 0.1–0.8 mm and 0.3–1.5 mm deep, respectively (Figure 10A–I). While it is difficult to establish a definite relationship between the parameters of Hybrid and depth of injury, it was observed that power increases depth of injury under APC mode.

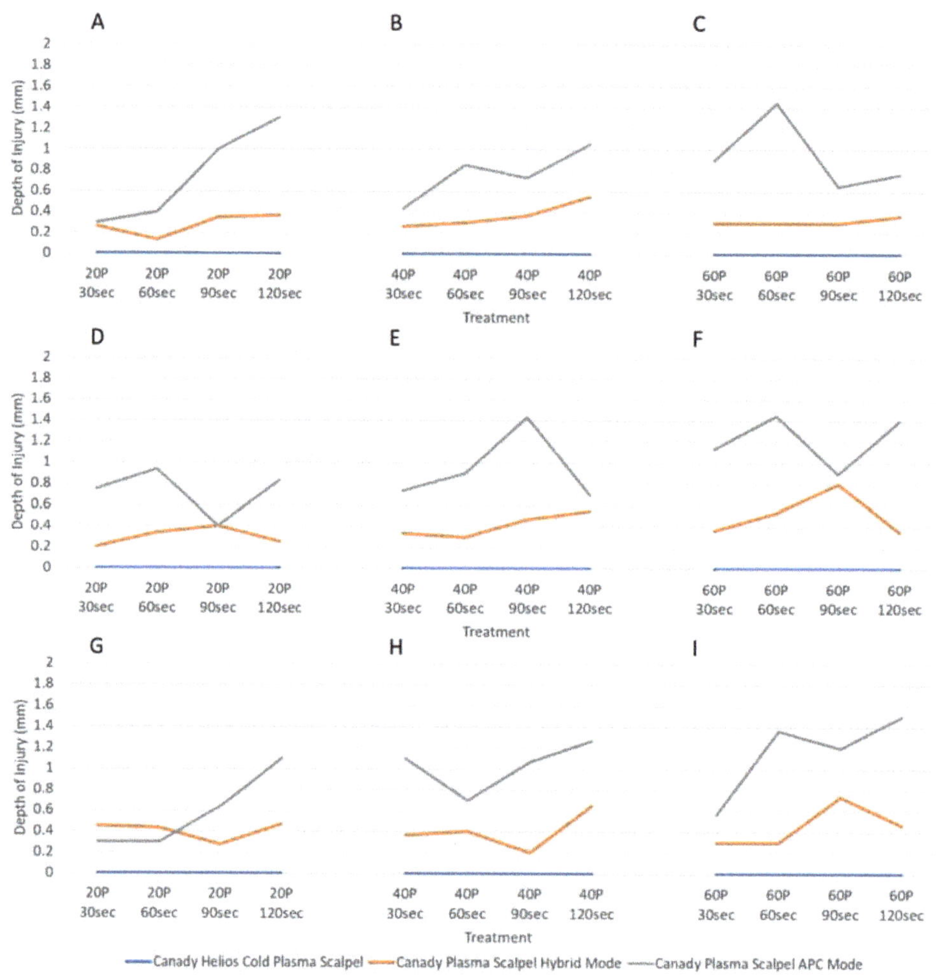

Figure 10. Depth of injury caused by each scalpel was measured post-treatment. Treatment conditions were 1 L/min (**A–C**), 3 L/min (**D–F**), and 5 L/min (**G–I**) flow rates with different powers and durations.

4. Discussion

This study explored the thermal injury between three plasma electrosurgical systems. Due to the differences in the operation and function of each electrosurgical system, a vast amount of data on inflicted tissue damage was collected. Severity of thermal injury, length of each plasma beam, depth of injury, and eschar (lateral spread) was compared between the Canady Plasma® Scalpels.

The Canady Helios Cold Plasma® Scalpel can generate CAP irrespective of the distance between plasma source and target tissue. For comparison purposes, plasma length was sustained at 2 cm which was typically longer than that of Hybrid and APC modes. It was observed that the CAP jet only attained tissue contact under higher flow rate conditions (3.0 and 5.0 L/min). Despite direct interaction between CAP and tissue in result of longer plasma columns, the eschar diameter and depth of injury remained relatively close to zero and tissue temperature change was minimal for all treatment conditions. The larger distance between the scalpel and the treatment area lessened the direct interaction of plasma with tissue and minimized collateral damage. It was demonstrated that the Canady Helios Cold Plasma® Scalpel induced no thermal damage to the normal liver tissue for all conditions tested.

The Canady Plasma® Scalpel in Hybrid and APC modes both depend on the ionization of argon gas but differ in the mechanism by which each plasma is produced. While Hybrid mode is associated with intermittent interruptions in a continuous low voltage waveform with high duty cycle, APC mode relies on spikes in a high voltage waveform with low duty cycle for coagulation. Since direct contact of scalpel and tissue is required for simultaneous cut and coagulation, plasma length could not be measured for Hybrid mode. In contrast, plasma length for APC mode was measurable as APC initiation is contact independent, like CAP mode.

Regarding tissue temperature changes, voltage is positively correlated to temperature. The comparatively lower tissue heating by Hybrid mode is a result of low voltage. Considering physical injury measurements alongside temperature measurements, Hybrid mode can potentially produce incisions and coagulation without causing excessive thermal damages.

The high voltage necessary for APC resulted in notably high temperatures which are ideal for sealing off blood vessels and ultimately preventing blood loss. Since tissue heating consequentially leads to thermal damage, APC produced a greater area of injury compared to Hybrid and CAP modes.

In general, adjustments to CAP parameters had little effect on plasma length, temperature, and injury measurements. For Hybrid mode, plasma length, temperature, and depth of injury also remained unaffected with an increase of plasma dosage. However, it was observed that injury width increased with power. For APC mode, power contributed to plasma length, injury diameter, and injury depth while higher flow rates and longer treatment times induced greater tissue heating.

5. Conclusions

Plasma length, temperature change, injury depth, and injury diameter were measured as results of the treatment with the Canady Plasma® Scalpels on different operational settings. Thermal damages varied amongst the three modes. Hybrid mode was found to produce moderate damages while APC mode demonstrated more severe tissue damages. The authors report a new Cold Plasma Jet System caused no thermal or structural damages to normal liver tissue. These findings were significantly less than that of Hybrid and APC modes. With the ability to treat non-cancerous tissue below the threshold of thermal damages, the Canady Helios Cold Plasma® Scalpel has immense potential in cancer therapy.

Supplementary Materials: Data for plasma length, tissue temperature change, eschar width, and depth of injury for all treatment conditions has been included as supplementary materials. Supplementary Materials are available online at http://www.mdpi.com/2571-6182/1/1/17/s1.

Author Contributions: Conceptualization, J.C. and M.K.; Data curation, S.J., T.N. and J.C.; Formal analysis, L.L., S.J., A.S., T.Z., S.W., T.N. and J.C.; Methodology, J.C.; Project administration, J.C.; Resources, J.C. and T.Z.; Writing—original draft, L.L.; Writing—review & editing, L.L., J.C., T.Z., W.R., X.C., M.K.

Funding: This research was funded by the US Medical Innovations.

Acknowledgments: The authors would like to thank the engineer team at Plasma Medicine Life Sciences for technical support of the plasma unit.

Conflicts of Interest: The authors declare no conflict of interest.

References

1. D'Arsonval, A. Action physiologique des courants alternatis a grande frequence. *Arch. Physiol. Porm. Pathol.* **1893**, *5*, 780–790.
2. Bovie, W.T. Electrosurgical Apparatus. U.S. Patent 1,813,902, 14 July 1931.
3. Cushing, H.; Bovie, W.T. Electrosurgery as an aid to the removal of intracranial tumors. In *Surgery, Gynecology and Obstetrics*; Surgical Publishing Company: Chicago, IL, USA, 1928.
4. Morrison, C.F., Jr. Electrosurgical Method and Apparatus for Initiating an Electrical Discharge in an Inert Gas Flow. U.S. Patent 40,404,426, 9 August 1977.

5.   Canady, J. Surgical Coagulation Device. U.S. Patent 5,207,675, 4 May 1993.

6.   Canady, J.; Vieira, E.; Vieira, N.; Wiley, K. System and Method for Electrosurgical Conductive Gas Cutting for Improving Eschar, Sealing Vessels and Tissues. U.S. Patent 2013/0296846, 2 November 2010.

7.   Gjika, E.; Pekker, M.; Shashurin, A.; Shneider, M.; Zhuang, T.; Canady, J.; Keidar, M. The cutting mechanism of the electrosurgical scalpel. *J. Phys. D Appl. Phys.* **2017**, *50*, 025401. [CrossRef]

8.   Gjika, E.; Scott, D.; Shashurin, A.; Zhuang, T.; Canady, J.; Keidar, M. Plasma-tissue interactions in argon plasma coagulation: Effects of power and tissue resistance. *Plasma Med.* **2016**, *6*, 125–134. [CrossRef]

9.   Canady, J.; Shashurin, A.; Wiley, K.; Fisch, N.J.; Keidar, M. Characterization of plasma parameters and tissue injury produced by plasma electrosurgical systems. *Plasma Med.* **2013**, *3*, 279–289. [CrossRef]

10.   Canady, J.; Shashurin, A.; Keidar, M.; Zhuang, T. Integrated Cold Plasma and High Frequency Plasma Electrosurgical System and Method. U.S. Patent 9,999,452, 19 June 2018.

11.   Keidar, M.; Shashurin, A.; Volotskova, O.; Ann Stepp, M.; Srinivasan, P.; Sandler, A.; Trink, B. Cold atmospheric plasma in cancer therapy. *Phys. Plasmas* **2013**, *20*, 057101. [CrossRef]

12.   Guerrero-Preston, R.; Ogawa, T.; Uemura, M.; Shumulinsky, G.; Valle, B.L.; Pirini, F.; Ravi, R.; Sidransky, D.; Keidar, M.; Trink, B. Cold atmospheric plasma treatment selectively targets head and neck squamous cell carcinoma cells. *Int. J. Mol. Med.* **2014**, *34*, 941–946. [CrossRef] [PubMed]

13.   Kim, S.J.; Chung, T.H. Cold atmospheric plasma jet-generated rons and their selective effects on normal and carcinoma cells. *Sci. Rep.* **2016**, *6*, 20332. [CrossRef] [PubMed]

14.   Keidar, M.; Walk, R.; Shashurin, A.; Srinivasan, P.; Sandler, A.; Dasgupta, S.; Ravi, R.; Guerrero-Preston, R.; Trink, B. Cold plasma selectivity and the possibility of a paradigm shift in cancer therapy. *Br. J. Cancer* **2011**, *105*, 1295–1301. [CrossRef] [PubMed]

15.   Kumar, N.; Park, J.H.; Jeon, S.N.; Park, B.S.; Choi, E.H.; Attri, P. The action of microsecond-pulsed plasmaactivated media on the inactivation of human lung cancer cells. *J. Phys. D App. Phys.* **2016**, *49*, 115401. [CrossRef]

16.   Kumar, N.; Attri, P.; Dewilde, S.; Bogaerts, A. Inactivation of human pancreatic ductal adenocarcinoma with atmospheric plasma treated media and water: A comparative study. *J. Phys. D App. Phys.* **2018**, *51*, 255401. [CrossRef]

17.   Rowe, W.; Cheng, X.; Ly, L.; Zhuang, T.; Basadonna, G.; Trink, B.; Keidar, M.; Canady, J. The Canady Helios Cold Plasma Scalpel Significantly Decreases Viability in Malignant Solid Tumor Cells in a Dose-Dependent Manner. *Plasma* **2018**. accepted. [CrossRef]

18.   Cheng, X.; Rowe, W.; Ly, L.; Shashurin, A.; Zhuang, T.; Wigh, S.; Basadonna, G.; Trink, B.; Keidar, M.; Canady, J. Treatment of Triple-negative Breast Cancer Cells with the Canady Cold Plasma Conversion System: Preliminary Results. *Plasma* **2018**. submitted.

 *plasma*

Article

Cold Atmospheric Pressure Plasma Treatment Modulates Human Monocytes/Macrophages Responsiveness

Letizia Crestale [1,†], **Romolo Laurita** [2,†], **Anna Liguori** [2,3,†], **Augusto Stancampiano** [2,†,‡], **Maria Talmon** [1,†], **Alina Bisag** [2], **Matteo Gherardi** [2,3], **Angela Amoruso** [4], **Vittorio Colombo** [2,5,*] and **Luigia G. Fresu** [1,*]

1 Department of Health Sciences, University of Piemonte Orientale, 28100 Novara, Italy; letiziacrest@gmail.com (L.C.); maria.talmon@med.uniupo.it (M.T.)
2 Department of Industrial Engineering, Alma Mater Studiorum-Università di Bologna, 40100 Bologna, Italy; romolo.laurita@unibo.it (R.L.); anna.liguori@unibo.it (A.L.); augusto.stancampiano@unibo.it (A.S.); alina.bisag@unibo.it (A.B.); matteo.gherardi4@unibo.it (M.G.)
3 Interdepartmental Centre for Industrial Research, Advanced Applications in Mechanical Engineering and Materials Technology, Alma Mater Studiorum-Università di Bologna, 40100 Bologna, Italy
4 R & D Biolab, 28100 Novara, Italy; angela.amoruso@gmail.com
5 Interdepartmental Centre for Industrial Research Agrifood, Alma Mater Studiorum-Università di Bologna, 47521 Cesena, Italy
* Correspondence: vittorio.colombo@unibo.it (V.C.); luigia.fresu@med.uniupo.it (L.G.F.)
† The authors have contributed equally.
‡ Current address: GREMI, UMR7344 CNRS/Université d'Orléans, 45067, Orléans, France.

Received: 30 September 2018; Accepted: 19 October 2018; Published: 29 October 2018

Abstract: Monocytes are involved in innate immune surveillance, establishment and resolution on inflammation, and can polarize versus M1 (pro-inflammatory) or M2 (anti-inflammatory) macrophages. The possibility to control and drive immune cells activity through plasma stimulation is therefore attractive. We focused on the effects induced by cold-atmospheric plasma on human primary monocytes and monocyte-derived macrophages. Monocytes resulted more susceptible than monocyte-derived macrophages to the plasma treatment as demonstrated by the increase in reactive oxygen (ROS) production and reduction of viability. Macrophages instead were not induced to produce ROS and presented a stable viability. Analysis of macrophage markers demonstrated a time-dependent decrease of the M1 population and a correspondent increase of M2 monocyte-derived macrophages (MDM). These findings suggest that plasma treatment may drive macrophage polarization towards an anti-inflammatory phenotype.

Keywords: cold atmospheric pressure plasma; dielectric barrier discharges; monocytes; monocytes-derived macrophages

---

1. Introduction

Monocytes are versatile mononuclear phagocytes involved in innate immune surveillance, establishment and resolution of inflammation. When activated, monocytes are recruited from the bloodstream to the site of inflammation where they differentiate into macrophages [1]. The local inflamed microenvironment drives polarization towards activated M1 macrophages, which display a pro-inflammatory phenotype and are involved in severe inflammation leading to tissue damage [2], or towards activated M2 macrophages [3], that display anti-inflammatory properties and are involved in tissue remodeling, wound healing, and efficient phagocytic activity [4]. M1 macrophages express high levels of major histocompatibility complex class II (MHC II) proteins, including the CD68 marker, and costimulatory molecules CD80 and CD86 [5]. They release reactive oxygen intermediates and several pro-inflammatory cytokines [6]. M2 macrophages are instead characterized by the expression of specific phenotypic markers, such as the mannose receptor-1 (CD206), the scavenger receptors CD163 and CD36 [7,8]. The molecular mechanisms underlying macrophage polarization have not been completely understood, because of the broad spectrum of stimuli affecting the process. However, it has been widely reported that this physiological process is altered in pathological conditions, for example in cancer or autoimmune diseases [9].

In this context, great interest has arisen towards physical stimuli that are able to modulate M1/M2 macrophage polarization [10]. In particular, the possibility to stimulate immune cells through cold atmospheric pressure plasma (CAP) treatment in order to control and drive their activity may pave the way to a vast field of medical applications. CAPs are characterized by a non-equilibrium in temperature between electrons and the heavy species of plasma (i.e., ions, excited species and neutrals) generated by the application of an electric field to a neutral gas. Besides presenting a negligible heat component, CAPs are characterized by several biological active components, such as ions, electrons, reactive oxygen (ROS) and nitrogen (RNS) species, UV radiation, playing a synergic action in the interaction of CAP with biological substrates [11]. Therefore, CAPs may be used for the treatment of cells and biological tissues by properly selecting the most suitable source and operating conditions for plasma generation to avoid any thermal damage [12,13]. Indeed, the Authors have previously shown an effect of CAP in a number of different eukaryotic cell types [14–17]. Immune cells have not been investigated yet thoroughly, although there are indications that their function may be affected. As an example, Kaushik et al. [18], reported the cytotoxic effects of CAP on monocytic lymphoma U937 cells and Bekeschus et al. [19] demonstrated the differential sensitivity of blood mononuclear cell subpopulations to plasma treatment. Focusing on the effects of plasma on monocytes, the plasma treatment of primary human monocytes can activate the pro-proliferative or pro-apoptotic intracellular signaling cascades, depending on plasma treatment time [20]. Several papers have focused on the plasma treatment of macrophagic populations [21–25]. In particular, the direct or indirect plasma treatment of macrophages can increase migration, an important immune cell function against diseases, and anti-tumor function [26,27].

In the present work, we focused our attention on the effects of a Dielectric Barrier Discharge (DBD) CAP source operated in open air on primary human monocytes and in monocyte-derived macrophages (MDM), evaluating their viability, ROS production and membrane markers expression.

2. Materials and Methods

2.1. Monocytes Isolation and Differentiation

The study was conducted in accordance with the Declaration of Helsinki. Ten healthy volunteers were enrolled, after approval of the Research Protocol by the Ethic Committee of Azienda Ospedaliera Maggiore della Carità, Novara, Italy (241CE), and informed written consent. Human monocytes were isolated from heparinised venous blood samples by standard technique of dextran sedimentation and histopaque (density = 1.077 $g \cdot cm^{-3}$, Sigma-Aldrich, St. Louis, MO, USA) gradient centrifugation ($400\times g$, 30 min, room temperature) and recovered by thin suction at the interface, as previously described [28]. Purified monocytes populations were obtained by adhesion (90 min, 37 °C, 5% CO_2) in serum free RPMI 1640 medium (Sigma-Aldrich) supplemented with 2 mM glutamine and antibiotics (Invitrogen, Carlsbad, CA, USA). After 90 min, medium was changed with RPMI added by 10% foetal bovine serum (FBS, Euroclone, Pero, Italy). Cell viability (trypan blue dye exclusion) was usually >98%. For differentiation into monocyte-derived macrophages (MDM), freshly isolated monocytes were cultured in 20% FBS-enriched medium for six days [29,30]. For plasma treatment cells were plated into 12-well plates in 1 mL of fresh medium.

2.2. Plasma Treatment

The CAP adopted in this study and reported in Figure 1A is a DBD source, already tested for biological applications [14,16,31,32] and consisting of a cylindrical brass electrode, 10 mm in diameter, having a hemispherical tip, with 5 mm of curvature radius. The electrode is coated with a 1 mm thick borosilicate glass (relative permittivity $\varepsilon_r = 4.7$), as dielectric layer. The DBD plasma source was operated in open air and powered by a micropulsed generator producing high-voltage quasi-sinusoidal pulses with peak voltage (PV) of 25 kV, pulse repetition frequency (f) of 20 kHz, and a duty cycle of 7.5%. In order to enable the plasma generation between the high voltage electrode and the liquid surface, the multiwell plate was positioned onto a grounded counter-electrode (aluminum foil with thickness of 0.13 mm). Plasma treatments were performed by setting the gap between the tip of the plasma source and the surface of the liquid medium at 2 mm. The experimental setup employed for the treatment is schematically reported in Figure 1B. In all the experiments, the temperature of the medium after CAP treatment resulted below the threshold of cytotoxicity. After the treatment cells were incubated 2 h at 37 °C, and 5% CO_2 before further analysis. The setup reported in Figure 1C was used for the time-resolved record of the plasma discharge electrical parameters. A high voltage probe (Tektronix P6015A, Tektronix, Beaverton, OR, USA) was used to measure the voltage waveform. The discharge current was measured by means of a current probe (Pearson 6585, Pearson Electronics, Palo Alto, CA, USA) mounted on the ground cable. Both signals were recorded with an oscilloscope (Tektronix DPO 40034, Tektronix) and subsequently elaborated to estimate the average power.

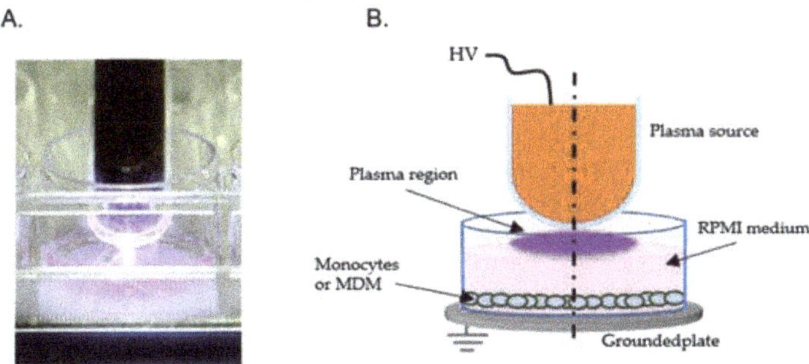

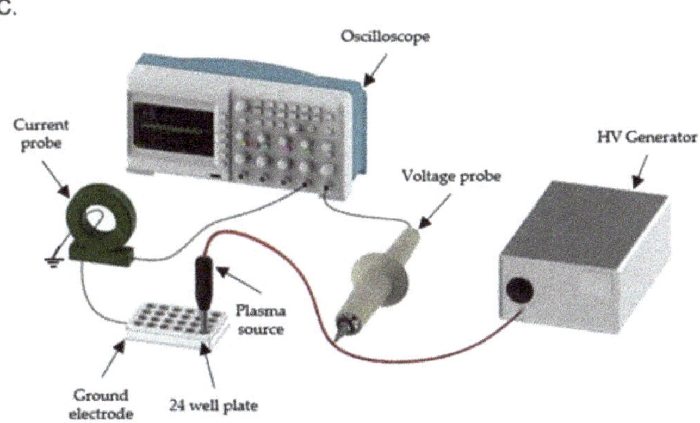

Figure 1. Dielectric Barrier Discharge (DBD) plasma source and setup. (**A**) Picture of plasma generated by a Dielectric Barrier Discharge source on cells; experimental setups employed for (**B**) the plasma treatment of monocytes and monocyte-derived macrophages and (**C**) the electrical characterization.

2.3. Detection of Reactive Oxygen and Nitrogen Species in Plasma-Treated Medium

The Amplex® Red Hydrogen Peroxide Assay Kit (Thermo Fisher Scientific, Waltham, MA, USA) and nitrate/nitrite colorimetric assay (ROCHE, Basel, Switzerland) were used, according to the manufacturer's protocol, to measure the concentrations of hydrogen peroxide and nitrites induced by plasma treatment in 1 mL of cell culture medium. Plasma treated medium was diluted 100 fold in phosphate buffered saline (PBS, a water-based salt solution containing 10 mM PO_4^{3-}, 137 mM NaCl, and 2.7 mM KCl, at pH 7.4) immediately after treatment. The absorbances were measured photometrically with a microplate reader (Rayto, Shenzhen, China).

2.4. Viability Test

To assess potential plasma toxicity in monocytes and MDM, cell viability was evaluated using the methylthiazolyldiphenyl-tetrazolium bromide (MTT) assay. Cells (1×10^5 cells) were treated with plasma in the above described conditions. Two hours after the treatment, the medium was replaced by the MTT assay solution (1 mg·mL^{-1}; 2 h, 37 °C 5% CO_2; Sigma-Aldrich). Supernatant was removed and DMSO (Sigma-Aldrich) was added in order to dissolve the purple formazan; the absorbance was measured at 580 and 675 nm. Treatment times were 5 s, 10 s and 20 s for monocytes, and 10 s, 20 s and 30 s for macrophages.

2.5. ROS/RNS and Superoxide Anion (O_2^-) Production

2 h after the treatment, the medium was changed and reactive species production was evaluated. O_2^- production was evaluated by the superoxide dismutase-sensitive cytochrome C reduction assay [33]. Briefly, 10^6 monocytes/macrophages treated by CAP were incubated with the cytochrome C (1 mg·mL^{-1}; 2 h, 37 °C 5% CO_2; Sigma-Aldrich) and then the supernatant was read at the spectrophotometer (Perkin Elmer Victor LightPerkin Elmer, MA, USA) at 550 nm. Indeed, cytochrome C reacting with the O_2^- is reduced in ferrocytochrome C whose absorbance is detectable at 550 nm. The results were expressed in nmol cytochrome C reduced/10^6 cells/30 min, using an extinction coefficient of 21.1 mm [34], and correlated with the amount of superoxide anion produced by analyzed cells. Moreover, ROS/RNS and O_2^- productions were also evaluated by flow cytometry analysis (FACS Calibur, BD, San Jose, CA, USA) using the Cellular ROS/Superoxide Detection Assay Kit (AbCam, Cambridge, UK) according to the manufacturer's instructions. The kit provides two fluorescent dye reagents: One (ROS/RNS, green) recognizing reactive species of both oxygen and nitrogen except for the superoxide anion that is detected by the second probe (O_2^-, Orange). Results were analyzed by Flowing Software (version 2.5; PerttuTerho, Turku Centre for Biotechnology, Turku, Finland) and expressed as percentage of cells expressing ROS/RNS or O_2^-. The cut-off for the analysis was based on non-stained sample. Moreover, we analyzed with the same software the mean fluorescence intensity (MFI) as indicator of the mean amount of O_2^- produced by each single cell.

2.6. Flow Cytometry Analysis

Evaluation of surface markers expression was performed by multi-parametric analysis by flow cytometry (FACS Calibur, BD) and analyzed by Flowing Software (version 2.5; PerttuTerho, Turku Centre for Biotechnology). Monocytes and MDM were treated by CAP and after 2 h cells were mechanically detached and stained for FACS analysis. The following antibody panels were used: Anti-CD14 (APC, eBioscience, MA, USA), anti-CD16 (FITC, eBioscience), anti-CD36 (FITC, eBioscience), anti-CD86 (PE, eBioscience), anti-CD163 (PE, eBioscience), and anti-CD206 (PerCp, eBioscience), with 10.000 events acquired. The monocytes and MDM population were defined as CD14$^+$ cells. Data were therefore expressed as the number of CD16$^+$, CD86$^+$, CD36$^+$, CD163$^+$ or CD206$^+$ cells over the number of CD14$^+$ cells (the gating strategy was shown in Figure 2). CD16 and CD86 are representative of M1 phenotype, while CD36, CD163 and CD206 of M2 phenotype.

Comparison between treated and untreated cells was performed and data were expressed as percentage of positive events.

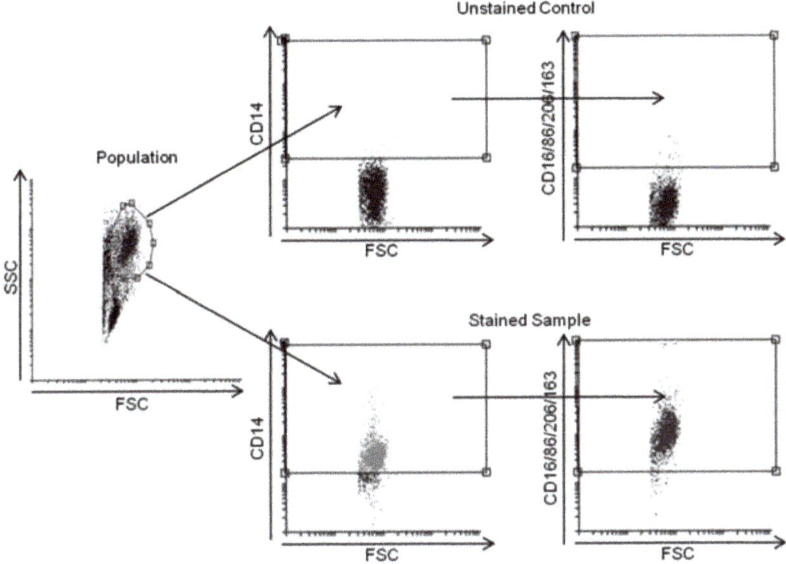

Figure 2. Gating strategy for flow cytometry (FACS) analysis of surface markers in monocytes and monocyte-derived macrophages (MDM). Population was first defined using forward scatter (FSC) and side scatter (SSC) to find viable cells and exclude debris. Then we gated CD14$^+$ cells and on this population we analyzed the expression of the other markers. All the gates were firstly set using unstained control.

2.7. Statistical Analysis

Statistical analyses were performed using GraphPad Prism 5. Data are presented as mean ± SEM (standard error of the mean) of "n" independent experiments performed in triplicate on monocytes/macrophages. Data were analyzed by one-way ANOVA non-parametric (Kruskal-Wallis and Dunn's test). A value of $p < 0.05$ was considered significant.

3. Results

3.1. Plasma Treatment and Viability Assay

In Figure 3 is reported the temporal evolution of voltage and current waveforms. Concerning the current waveform is possible to observe multiple peaks both in the positive and negative half-periods of the voltage pulse that may be associated with multiple discharge events. The subsequently processing of waveforms of three independent experiments allowed to estimate an average power density of 1.7 ± 0.1 W.

In Figure 4, the concentrations of hydrogen peroxides and nitrites in 1 mL of culture media after plasma treatments are reported. CAP treatment induced the production of similar concentrations of nitrites (up to about 340 µM after 60 s) in both monocytes and MDM culture medium. On the other hand, hydrogen peroxide concentrations in monocyte culture medium were significantly higher compared to those produced in MDM culture medium.

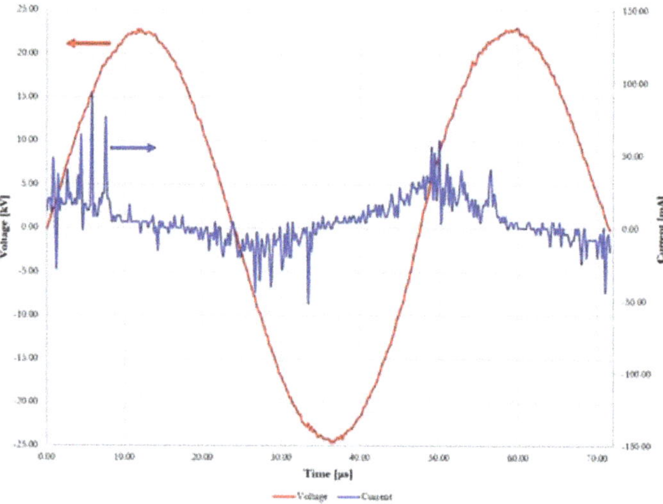

Figure 3. Typical voltage (blue) and current (red) waveforms recorded during the plasma treatment.

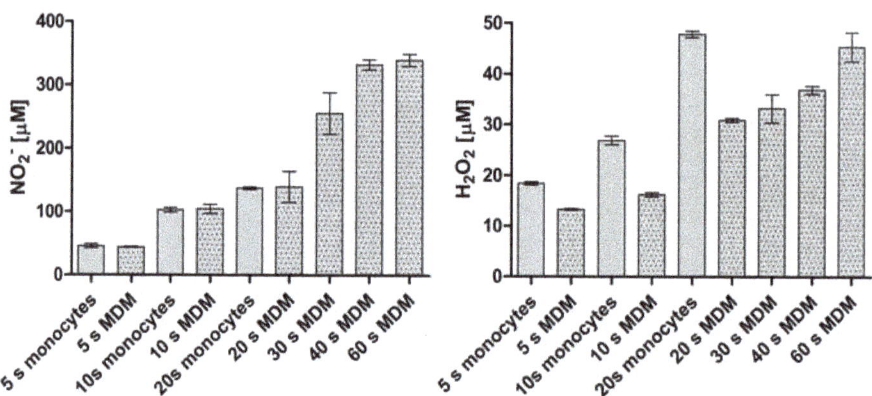

Figure 4. Nitrite and hydrogen peroxide concentrations in plasma treated media. Data are expressed as mean $\pm$ SEM of three independent experiments ($n = 9$).

Plasma exposure times were selected by evaluating the cytotoxicity induced by CAP in both cell models. Monocytes resulted to be very sensitive to CAP exposure for periods longer than 20 s (data not shown). Therefore, as reported in Figure 5A, the range of exposure times were limited to 5–20 s, despite a significant reduction in cell viability was observed after only 10 s. MDM were at first treated for 20 s, 40 s and 60 s, highlighting a strong reduction of viability after 40 s and 60 s of plasma exposure (data not shown). Accordingly, treatment times of 10 s, 20 s, and 30 s were then performed. In these operating conditions, CAP did not affect MDM viability (Figure 5B), with only a non-significant reduction (20%) of cell viability being registered after 30 s of plasma treatment.

A.

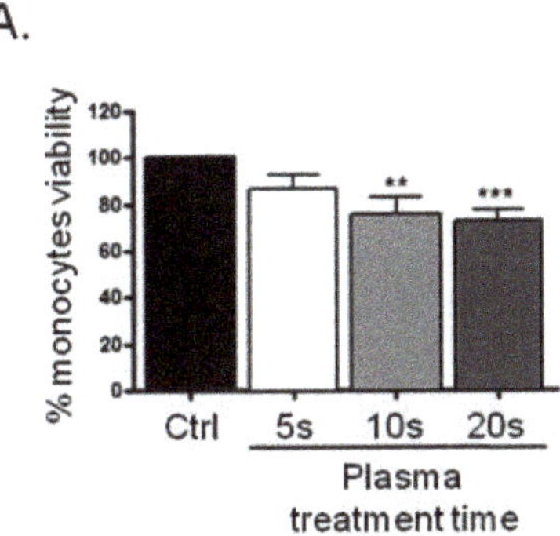

B.

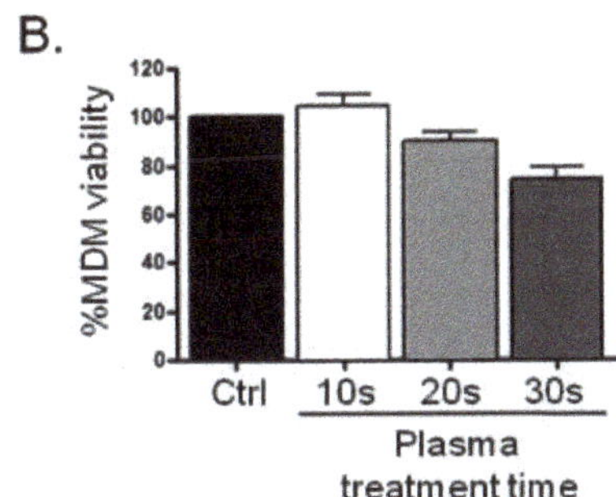

Figure 5. Cell viability after cold atmospheric pressure plasma (CAP) treatment. (**A**) Monocytes treated with CAP for 5 s, 10 s, and 20 s showed a significant reduction of viability already after 10 s of treatment. Data are expressed as mean ± SEM analyzed by Kruskal-Wallis test of 10 independent experiments ($n = 30$). ** $p < 0.005$, *** $p < 0.001$ vs. untreated control (Ctrl). (**B**) MDM treated by CAP for 10 s, 20 s, and 30 s.

3.2. Plasma Treatment Effects on Monocytes

3.2.1. ROS Production by Treated Monocytes

Since human monocytes are phagocytes and release oxy-radicals upon challenge with appropriate stimuli, we first investigated CAP ability to affect superoxide anion (O_2^-) production, using both an indirect and a direct method (superoxide dismutase-sensitive cytochrome C reduction assay and flow cytometry analysis, respectively), as reported in Materials and Methods. Basal O_2^- production of monocytes amounted to 0.46 ± 0.018 nmol cytochrome C reduced/10^6 cells, while after 20 s of CAP treatment, this was significantly increased by about 0.15 nmol (Figure 6A). This small, but significant, increase might be a direct consequence of the reduced viability observed. The cytofluorimetric assay confirmed this result demonstrating that the percentage of cells producing radical oxygen species moved from about $19 \pm 4\%$ of untreated cells to $47 \pm 1\%$ after 20 s of CAP treatment (Figure 6B). Moreover, the analysis of MFI, that in this instance represents the amount of superoxide anion produced by cells, revealed a progressive increase (Figure 6C) that become significant at 10 s, confirming that CAP treatment induced not only the oxidative burst in a greater number of cells, but also each cell to produce a higher amount of O_2^-. PMA (positive control of oxidative burst induction) 10^{-6} M was used as positive control of oxidative burst induction [35].

ROS/RNS evaluation in monocytes after CAP treatment could not be performed, since the percentage of expressing cells in the untreated group in our test conditions was high and therefore poorly responsive to treatments (untreated monocytes 79%±0.7 vs. monocytes treated with PMA 88%±2).

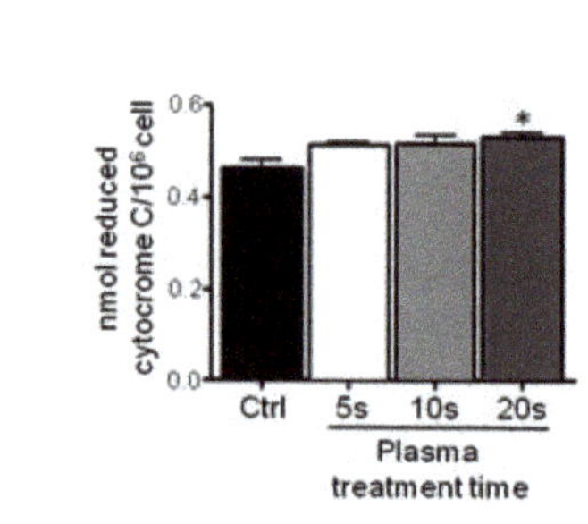

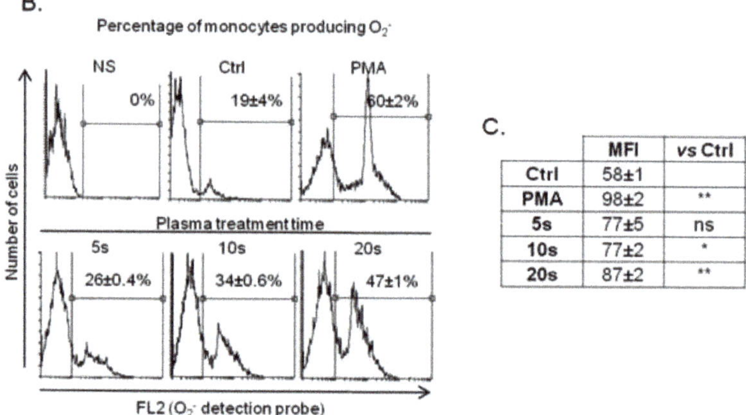

Figure 6. Effect of plasma treatment on monocytes superoxide anion production. (**A**) Monocytes were exposed to CAP for 5 s, 10 s and 20 s and results are expressed as levels of nmol reduced cytocromeC/10^6 cells. Data are expressed as mean ± SEM of five independent experiments analyzed by Kruskal-Wallis test and Dunn's test for multiple comparisons. * $p < 0.05$ vs. untreated cells (Ctrl); (**B**) Cytofluorimetric assay to assess positive monocytes to the superoxide anion staining. NS, unstained control used to set acquisition parameters; Ctrl, untreated control; PMA, positive control of oxidative burst induction. Results are expressed as percentage mean ± SEM of three independent experiments analyzed by Kruskal-Wallis test and Dunn's test for multiple comparisons. (**C**) Mean fluorescence intensity (MFI) analysis expressed as mean ± SEM of three independent experiments analyzed by Kruskal-Wallis test and Dunn's test for multiple comparison. * $p < 0.05$; ** $p < 0.005$ vs. corresponding control.

3.2.2. Surface Marker Expression

The expression of specific monocyte surface markers was then analyzed (Figure 6). Freshly isolated monocytes treated with CAP showed a significant reduction (about 40%) of CD86, CD36, CD163 and CD206 already after 5 s of treatment, (Figure 7) while the reduction of CD16 expression became significant after 10 s of plasma exposure.

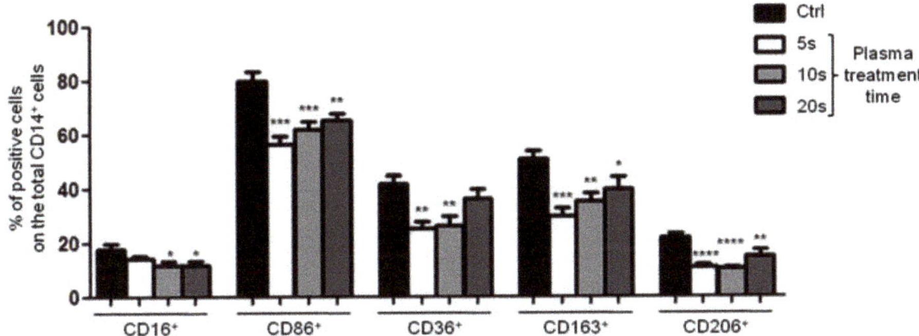

Figure 7. Effect of plasma treatment on membrane markers expression of monocytes. $CD14^+$ cells, treated and untreated (Ctrl), were stained with the indicated antibodies and analyzed by flow cytometry. Results are expressed as the percentage of positive events for each marker on the total of $CD14^+$. Data are expressed as mean ± SEM of 10 independent experiments analyzed by Kruskal-Wallis test and Dunn's test for multiple comparison. * $p < 0.05$; ** $p < 0.005$; *** $p < 0.001$; **** $p < 0.0001$ vs. corresponding control. We have previously shown that untreated monocytes do not change marker expression in a two-hour incubation and therefore the untreated control also depicts the baseline expression [36].

3.3. Plasma Treatment Effects on Monocytes-Derived Macrophages (MDM)

3.3.1. ROS Production by MDM

The production of both O_2^- and other reactive oxygen and nitrogen species (ROS/RNS) in MDM exposed to CAP is shown in Figure 6. MDM had a basal production of O_2^- of about a 0.39 ± 0.04 nmol of cytochrome C reduced/10^6 cells (Figure 8A), and CAP treatments of 10, 20 and 30 s were ineffective in inducing an oxidative burst, as demonstrated by the unchanged percentage of positive cells to the ROS/RNS staining (Figure 8B). In contrast, the number of positive MDM to the O_2^- staining was significantly increased after the longest time treatment (Figure 8C). It is interesting to note that after plasma treatment we can observe the onset of a peak of low-expressing O_2^- cells (represented by the circled peak in the histogram), that gradually increases in parallel with CAP-treatment time. This can be correlated with the emergence of a new population of MDM (in treated cells vs. Ctrl) able to produce basal levels of O_2^-. Moreover, it is important to note that the MFI did not increase (Figure 8D).

3.3.2. Surface Marker Expression

As shown in Figure 9A, in MDM exposed to CAP the expression of the single surface markers was not significantly influenced by CAP treatment, except for CD36, whose expression significantly increased, and CD16, that decreased already after 10 s. This result is in line with the previously presented data: In fact, CAP treatment did not induce the MDM respiratory burst nor a decrease in cell survival even after the longest treatment time. The possibility that CAP exposure could drive polarization of MDM versus M1 or M2 phenotypes was investigated. In fact, freshly isolated monocytes, cultured in 20% FBS-enriched medium, spontaneously differentiate to MDM, defined as M0, an intermediate phenotype (not M1 nor M2), that under different stimuli can shift to one phenotype or to the other. We performed a co-expression analysis for the specific markers of M1 (CD16/CD86) and M2 (CD163/206) on the total of $CD14^+$ population. As demonstrated in Figure 9B, there was a time dependent reduction of M1 population, defined as $CD14^+CD16^+CD86^+$, becoming significant at 30 s, with a correspondent increase to the M2 population, defined as $CD14^+CD163^+CD206^+$.

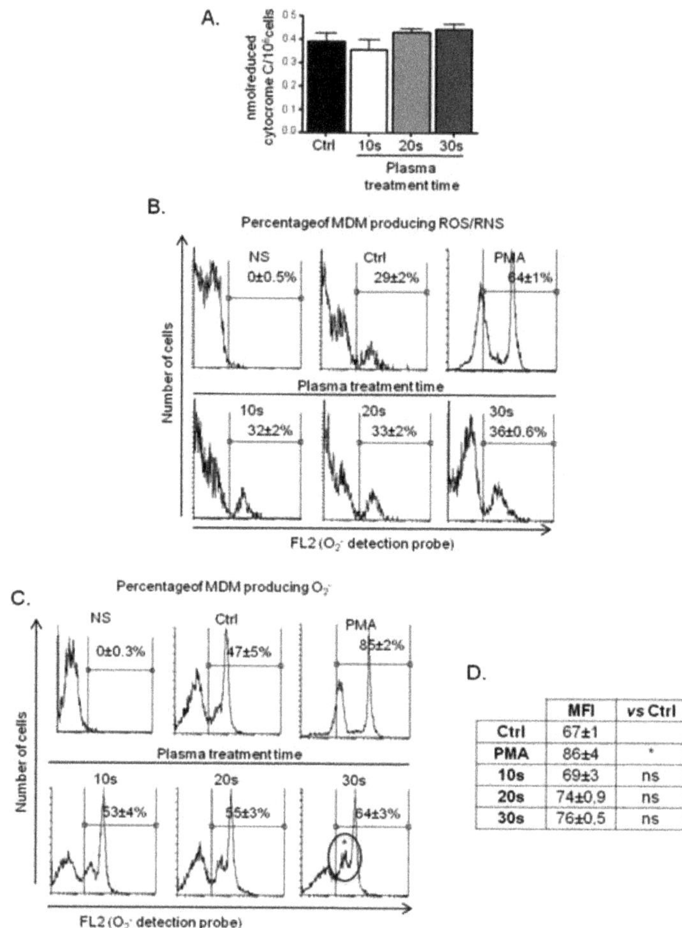

	MFI	*vs* Ctrl
Ctrl	67±1	
PMA	86±4	*
10s	69±3	ns
20s	74±0.9	ns
30s	76±0.5	ns

Figure 8. Effect of plasma treatment on reactive species production in MDM. (**A**) Superoxide anion production. Cells were irradiated by CAP for 10, 20 and 30 s, and results are expressed as levels of nmol reduced cytocromeC/10^6 MDM. Data are expressed as mean of five independent experiments analyzed by Kruskal-Wallis test and Dunn's test for multiple comparison; (**B**) Cytofluorimetric assay to assess MDM positive to the reactive oxygen (ROS)/RNS detection probe staining after 10, 20 and 30 s of treatment with CAP. Results are expressed as percentage mean of MDM producing ROS/RNS $\pm$ SEM of three independent experiments analyzed by Kruskal-Wallis test and Dunn's test for multiple comparison; (**C**) Cytofluorimetric assay to assess MDM positive to the superoxide anion detection probe staining after 10, 20 and 30 s of treatment with CAP. Results are expressed as percentage mean of MDM producing O_2^- $\pm$ SEM of three independent experiments analyzed by Kruskal-Wallis test and Dunn's test for multiple comparison. * $p < 0.05$ vs. untreated cell (Ctrl); (**D**) MFI analysis expressed as mean $\pm$ SEM of three independent experiments analyzed by Kruskal-Wallis test and Dunn's test for multiple comparisons. * $p < 0.05$ vs. corresponding control. NS, unstained control used to set acquisition parameters, Ctrl, untreated control; PMA, positive control of oxidative burst induction.

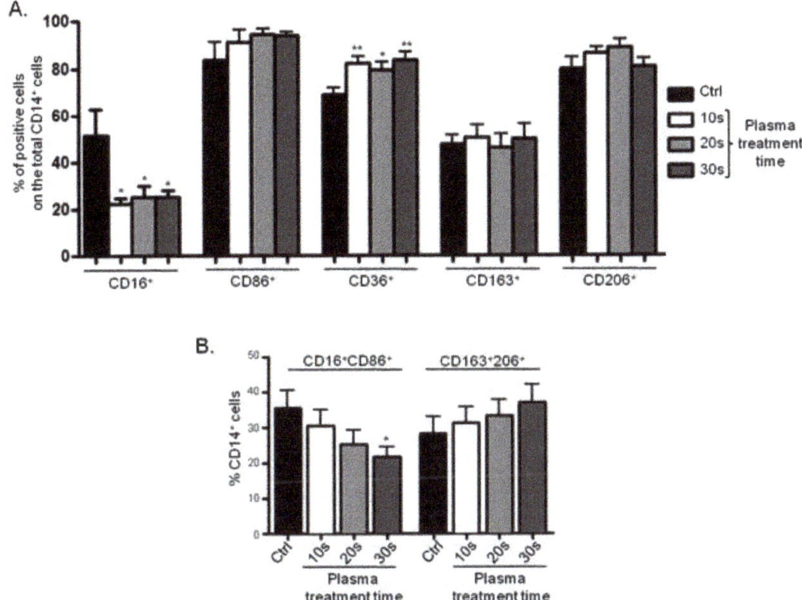

Figure 9. Effect of plasma treatment on MDM phenotype. (**A**) Cyofluorimetric analysis of MDM stained with anti-CD14, anti-CD86, anti-CD36, anti-CD163 and anti-CD206. Results are expressed as the percentage mean of positive events for each marker on the total of CD14$^+$ $\pm$ SEM of 10 independent experiments analyzed by Kruskal-Wallis test and Dunn's test for multiple comparison. * $p < 0.05$, ** $p < 0.005$ vs. untreated control (Ctrl); (**B**) Cytofluorimetric analysis of M1 and M2 populations represented by the co-expression of CD14/CD16/CD86 and CD14/CD163/CD206 respectively. * $p < 0.05$ vs. untreated cell (Ctrl). Data are expressed as mean $\pm$ SEM of 10 independent experiments analyzed by Kruskal-Wallis test and Dunn's test for multiple comparisons. We have previously shown that untreated MDM do not change marker expression in a two-hour incubation and therefore the untreated control also depicts the baseline expression.

4. Discussion

In this study, the effect of a microsecond pulsed DBD on human monocytes and monocyte-derived macrophages was investigated.

The CAP treatment of monocytes and MDM culture medium induced the production of nitrites and hydrogen peroxides. The lower the concentration of FBS, the higher the concentration of hydrogen peroxide, while the concentration of nitrites resulted similar in both treated media. This difference can be ascribed to the differences between MDM and monocytes culture medium in terms of FBS concentrations, scavenger of hydrogen peroxide, but not of nitrites [37]. Moreover, the longer the treatment time, the higher was the concentration of both species in the treated media and the reduction of cell viability for both monocytes and MDM. Similar findings have been reported for several cell lines [16,20,38].

Focusing on monocytes, CAP treatment up to 20 s induced cells to produce of O_2^- and a decrease of surface markers. This effect is unlikely to be due to a change in cell polarization, but is most likely a physiochemical modification, since compensatory increases in cell surface markers were not observed to counteract these decreases. While we did not investigate this further, the decrease of marker expression might be explained by a direct membrane damage (for example peroxidation and subsequent loss of membrane fluidity and elasticity [39]), to a direct protein oxidation [40] or by other mechanisms, which we have nonetheless not evaluated further. Anyway, it is noteworthy to highlight a slight recovery of the surface marker expression after the impressive 5 s reduction. This could be explained with a rapid anti-oxidative response of cell to the insult [41], but this defense mechanism was not sufficient to restore membrane integrity and keep cell survival. The fact that monocytes are significantly more sensitive to CAP compared to MDM would be in accordance with Bundscherer et al. [42] who demonstrated significant differences between different immune cell types regarding survival after plasma treatment.

On the other hand, CAP treatment for up to 30 s of MDM did not cause an increase in $O_2{}^-$ production, but only induced a greater number of cells to produce the superoxide anion. As a possible consequence, MDM, terminally differentiated cells, resulted more resistant to CAP treatment. Indeed, CAP treatment did not affect significantly MDM survival. We suggest that these results are strictly connected to the absence of an increase of ROS/RNS and the augmented percentage of $O_2{}^-$-producing MDM. In fact, as demonstrated by Zhang et al. [43] the polarization versus M2 phenotype is sustained by a proper amount of $O_2{}^-$. Indeed, when monocytes are induced to differentiate there is an increased superoxide anion production that triggers the biphasic ERK activation that has a pivotal role for M2 differentiation. In our experiments, CAP stimulated more M0 to secrete basal amounts of superoxide necessary to polarize versus M2, that were represented by the new population previously described in Figure 8C. Moreover, marker expression modifications in MDM were reputed as a true sign of polarization change, since CD16 decrease was counter-balanced by an increase in CD36, while CD163 and CD206 were not affected. In the present study, M2 subsets most represented in CAP treated MDM have not been evaluated. In the future, it will be interesting to explore this aspect as the different M2 sub-populations are involved in different physiological and pathological processes [4,44,45]. This information will therefore provide clues on the clinical applications of CAP.

5. Conclusions

In conclusion, our results suggest that CAP treatment may be able to selectively modulate the effect of some cells over others. Furthermore, CAP could drive macrophage polarization supporting the idea that at the proper operating conditions of plasma treatment it could be possible to direct macrophages versus an anti-inflammatory phenotype. At the same time, our results also suggest the differences in terms of viability of macrophages compared to monocytes after CAP exposure may be due to the different concentration of FBS between the cell culture media. In fact, as reported by Yan et al. [46], it is possible to harness the medium to kill glioblastoma cells by altering the concentration of the serum demonstrating therefore the significant role of media in CAP treatment. In prospect, our results may suggest that CAP technology may find clinical applications in those pathological settings in which macrophages play a prominent role, and in which it can be delivered locally, for example in wound healing, treatment of non-metastatic melanoma difficult to surgically remove or in topical applications for autoimmune disorders, such as psoriasis.

We acknowledge that our results should be read in light of the following limits: (a) We have evaluated changes in monocytes and MDM 2 h after CAP treatment, but events occurring immediately after treatment might be relevant; (b) we have not fully characterized the M2 sub-populations represented in CAP treated cells, neither by FACS nor by real time PCR. This last point, in the future, will yield more precise indications on the possible therapeutic applications of CAP.

Author Contributions: L.C., A.A., M.T., R.L., A.L., A.S., A.B.: methodology and investigation A.A., V.C., M.G.: conceptualization, L.G.F., R.L., A.L., A.S.: writing—original draft preparation, L.G.F., M.G., V.C.: supervision, formal analysis, M.G., V.G, L.G.F. Writing—review & editing.

Funding: This research received no external funding.

Conflicts of Interest: The authors declare no conflict of interest.

References

1. Shi, C.; Pamer, E.G. Monocyte recruitment during infection and inflammation. *Nat. Rev. Immunol.* **2011**. [CrossRef] [PubMed]
2. Murray, P.J.; Wynn, T.A. Protective and pathogenic functions of macrophage subsets. *Nat. Rev. Immunol.* **2011**, *11*, 723. [CrossRef] [PubMed]
3. Gordon, S.; Martinez, F.O. Alternative activation of macrophages: mechanism and functions. *Immunity* **2010**. [CrossRef] [PubMed]
4. Roszer, T. Understanding the Mysterious M2 Macrophage through Activation Markers and Effector Mechanisms. *Mediat. Inflamm.* **2015**. [CrossRef] [PubMed]
5. Cybulsky, M.I.; Cheong, C.; Robbins, C.S. Macrophages and Dendritic Cells: Partners in Atherogenesis. *Circ Res.* **2016**, *118*, 637. [CrossRef] [PubMed]
6. Das, A.; Sinha, M.; Datta, S.; Abas, M.; Chaffee, S.; Sen, C.K.; Roy, S. Monocyte and macrophage plasticity in tissue repair and regeneration. *Am. J. Pathol.* **2015**. [CrossRef] [PubMed]
7. Mantovani, A.; Biswas, S.K.; Galdiero, M.R.; Sica, A.; Locati, M. Macrophage plasticity and polarization in tissue repair and remodelling. *J. Path.* **2013**. [CrossRef] [PubMed]
8. Wynn, T.A.; Chawla, A.; Pollard, J.W. Macrophage biology in development, homeostasis and disease. *Nature* **2013**. [CrossRef] [PubMed]
9. Sica, A.; Mantovani, A. Macrophage plasticity and polarization: In vivo veritas. *J. Clin. Investig.* **2012**. [CrossRef] [PubMed]
10. Zhou, D.; Huang, C.; Lin, Z.; Zhan, S.; Kong, L.; Fang, C.; Li, J. Macrophage polarization and function with emphasis on the evolving roles of coordinated regulation of cellular signaling pathways. *Cell Sign.* **2014**, *2*, 192–197. [CrossRef] [PubMed]
11. Laroussi, M.M. Low Temperature Plasma-Based Sterilization: Overview and State-of-the-Art. *Plasma Process. Polym.* **2005**. [CrossRef]
12. Bekeschus, S.; Kolata, J.; Winterbourn, C.; Kramer, A.; Turner, R.; Weltmann, K.D.; Bröker, B.; Masur, K. Hydrogen peroxide: A central player in physical plasma-induced oxidative stress in human blood cells. *Free Radicals Res.* **2014**. [CrossRef] [PubMed]
13. Barbieri, D.B.; Cavrini, F.; Colombo, V.; Gherardi, M.; Landini, M.P.; Laurita, R.; Liguori, A.; Stancampiano, A. Investigation of the antimicrobial activity at safe levels for eukaryotic cells of a low power atmospheric pressure inductively coupled plasma source. *Biointerphases* **2015**. [CrossRef] [PubMed]
14. Gherardi, M.; Turrini, E.; Laurita, R.; De Gianni, E.; Ferruzzi, L.; Liguori, A.; Stancampiano, A.; Colombo, V.; Fimognari, C. Atmospheric Non-Equilibrium Plasma Promotes Cell Death and Cell-Cycle Arrest in a Lymphoma Cell Line. *Plasma Process. Polym.* **2015**. [CrossRef]
15. Laroussi, M.; Xi, L. Room-temperature atmospheric pressure plasma plume for biomedical applications. *Appl. Phys. Lett.* **2005**. [CrossRef]
16. Laurita, R.; Alviano, F.; Marchionni, C.; Abruzzo, P.M.; Bolotta, A.; Bonsi, L.; Colombo, V.; Gherardi, M.; Liguori, A.; Ricci, F.; et al. A study of the effect on human mesenchymal stem cells of an atmospheric pressure plasma source driven by different voltage waveforms. *J. Phys. D Appl. Phys.* **2016**. [CrossRef]
17. Stoffels, E.; Sakiyama, Y.; Graves, D.B. Cold Atmospheric Plasma: Charged Species and Their Interactions With Cells and Tissues. *IEEE Trans. Plasma Sci.* **2008**.
18. Kaushik, N.; Kumar, N.; Kim, C.H.; Kaushik, N.K.; Choi, E.H. Dielectric Barrier Discharge Plasma Efficiently Delivers an Apoptotic Response in Human Monocytic Lymphoma. *Plasma Process. Polym.* **2014**. [CrossRef]
19. Bekeschus, S.; Kolata, J.; Muller, A.; Kramer, A.; Weltmann, K.D.; Broker, B.; Masur, K. Differential Viability of Eight Human Blood Mononuclear Cell Subpopulations after Plasma Treatment. *Plasma Med.* **2013**. [CrossRef]
20. Bundscherer, L.; Nagel, S.; Hasse, S.; Tresp, H.; Wende, K.; Walther, R.; Lindequist, U. Non-thermal plasma treatment induces MAPK signaling in human monocytes. *Open Chem.* **2015**. [CrossRef]
21. Kaushik, N.; Kaushik, N.; Min, B.; Choi, K.; Hong, Y.; Miller, V.; Fridman, A; Choi, E.H. Cytotoxic macrophage-released tumour necrosis factor-alpha (TNF-α) as a killing mechanism for cancer cell death after cold plasma activation. *J. Phys. D Appl. Phys.* **2016**. [CrossRef]

22. Lin, A.; Truong, B.; Pappas, A.; Kirifides, L.; Oubarri, A.; Chen, S.; Lin, S.; Dobrynin, D.; Fridman, G; Fridman, A.; et al. Uniform Nanosecond Pulsed Dielectric Barrier Discharge Plasma Enhances Anti-Tumor Effects by Induction of Immunogenic Cell Death in Tumors and Stimulation of Macrophages. *Plasma Process. Polym.* **2015**. [CrossRef]

23. Miller, V.; Lin, A.; Fridman, G.; Dobrynin, D.; Fridman, A. Plasma Stimulation of Migration of Macrophages. *Plasma Process. Polym.* **2014**. [CrossRef]

24. Miller, V.; Lin, A.; Fridman, A. Why target immune cells for plasma treatment of cancer. *Plasma Chem. Plasma Process.* **2016**. [CrossRef]

25. Georgescu, N.; Lupu, R. Tumoral and normal cells treatment with high-voltage pulsed cold atmospheric plasma jets. *IEEE Trans. Plasma Sci.* **2010**. [CrossRef]

26. Lin, A.; Truong, B.; Fridman, G.; Fridman, A.; Miller, V. Immune cells enhance selectivity of nanosecond-pulsed dbd plasma against tumor cells. *Plasma Med.* **2017**. [CrossRef]

27. Bekeschus, S.; Scherwietes, L.; Freund, E.; Rouven Liedtke, K.; Hackbarth, C.; von Woedtke, T.; Partecke, L.I. Plasma-treated medium tunes the inflammatory profile in murine bone marrow-derived macrophages. *Clin. Plasma Med.* **2018**. [CrossRef]

28. Lavagno, L.; Gunella, G.; Bardelli, C.; Spina, S.; Fresu, L.G.; Viano, I.; Brunelleschi, S. Anti-inflammatory drugs and tumor necrosis factor-alpha production from monocytes: Role of transcription factor NF-kappa B and implication for rheumatoid arthritis therapy. *Eur. J. Pharmacol.* **2004**. [CrossRef] [PubMed]

29. Gantner, F.; Kupferschmidt, R.; Schudt, C.; Wendel, A.; Hatzelmann, A. In vitro differentiation of human monocytes to macrophages: Change of PDE profile and its relationship to suppression of tumour necrosis factor-alpha release by PDE inhibitors. *Br. J. Pharmacol.* **1997**. [CrossRef] [PubMed]

30. Amoruso, A.; Bardelli, C.; Gunella, G.; Fresu, L.G.; Ferrero, V.; Brunelleschi, S. Quantification of PPAR-gamma protein in monocyte/macrophages from healthy smokers and non-smokers: A possible direct effect of nicotine. *Life Sci.* **2007**. [CrossRef] [PubMed]

31. Babington, P.; Rajjoub, K.; Canady, J.; Siu, A.; Keidar, M.; Sherman, J.H. Use of cold atmospheric plasma in the treatment of cancer. *Biointerphases* **2015**. [CrossRef] [PubMed]

32. Turrini, E.; Laurita, R.; Stancampiano, A.; Catanzaro, E.; Calcabrini, C.; Maffei, F.; Gherardi, M.; Colombo, V.; Fimognari, C. Cold atmospheric plasma induces apoptosis and oxidative stress pathway regulation in T-lymphoblastoid leukemia cells. *Oxid. Med. Cell. Longevity* **2017**. [CrossRef] [PubMed]

33. Bardelli, C.; Gunella, G.; Varsaldi, F.; Balbo, P.; Del Boca, E.; Bernardone, I.S.; Amosuso, A.; Brunelleschi, S. Expression of functional NK1 receptors in human alveolar macrophages: Superoxide anion production, cytokine release and involvement of NF-κB pathway. *Br. J. Pharmacol.* **2005**. [CrossRef] [PubMed]

34. Van Gelden, B.; Slater, E.C. The extinction coefficient of cytochrome c. *Biochim. Biophys. Acta* **1962**, *58*, 593. [CrossRef]

35. Myers, M.A.; McPhail, L.C.; Snyderman, R. Redistribution of protein kinase C activity in human monocytes: Correlation with activation of the respiratory burst. *J Immunol.* **1985**, *135*, 3411. [PubMed]

36. Talmon, M.; Rossi, S.; Pastore, A.; Cattaneo, C.I.; Brunelleschi, S.; Fresu, L.G. Vortioxetine exerts anti-inflammatory and immunomodulatory effects on human monocytes/macrophages. *Br. J. Pharmacol.* **2018**. [CrossRef] [PubMed]

37. Boehm, D.; Heslin, C.; Cullen, P.J.; Bourke, P. Cytotoxic and mutagenic potential of solutions exposed to cold atmospheric plasma. *Sci. Rep.* **2016**, *6*, 21464. [CrossRef] [PubMed]

38. Kalghatgi, S.; Friedman, G.; Fridman, A.; Clyne, A.M. Endothelial cell proliferation is enhanced by low dose non-thermal plasma through fibroblast growth factor-2 release. *Ann. Biomed. Eng.* 2010. [CrossRef] [PubMed]

39. Wong-Ekkabut, J.; Xu, Z.; Triampo, W.; Tang, I.M.; Tieleman, D.P.; Monticelli, L. Effect of lipid peroxidation on the properties of lipid bilayers: A molecular dynamics study. *Biophys. J.* **2007**. [CrossRef] [PubMed]

40. Davies, M.J. Protein oxidation and peroxidation. *Biochem. J.* **2016**. [CrossRef] [PubMed]

41. Baran, C.P.; Zeigler, M.M.; Tridandapani, S.; Marsh, C.B. The role of ROS and RNS in regulating life and death of blood monocytes. *Curr. Pharm. Des.* **2004**. [CrossRef]

42. Bundscherer, L.; Bekeschus, S.; Tresp, H.; Hasse, S.; Reuter, S.; Weltmann, K.; Lindequist, U.; Masur, K. Viability of Human Blood Leukocytes Compared with Their Respective Cell Lines after Plasma Treatment. *Plasma Med.* **2013**. [CrossRef]

43. Zhang, Y.; Choksi, S.; Chen, K.; Pobezinskaya, Y.; Linnoila, I.; Liu, Z.G. ROS play a critical role in the differentiation of alternatively activated macrophages and the occurrence of tumor-associated macrophages. *Cell Res.* **2013**. [CrossRef] [PubMed]

44. Mantovani, A.; Sica, A.; Sozzani, S.; Allavena, P.; Vecchi, A.; Locati, M. The chemokine system in diverse forms of macrophage activation and polarization. *Trends Immunol.* **2004**. [CrossRef] [PubMed]

45. Duluc, D.; Delneste, Y.; Tan, F.; Moles, M.P.; Grimaud, L.; Lenoir, J.; Preisser, L.; Anegon, I.; Catala, L.; Ifrah, N.; et al. Tumor-associated leukemia inhibitory factor and IL-6 skew monocyte differentiation into tumor-associated macrophage-like cells. *Blood* **2007**. [CrossRef] [PubMed]

46. Yan, D.; Sherman, J.H.; Cheng, X.; Ratovitski, E.; Canady, J.; Keidar, M. Controlling plasma stimulated media in cancer treatment application. *Appl. Phys. Lett.* **2014**. [CrossRef]

Article

Possible Mechanism of Glucose Uptake Enhanced by Cold Atmospheric Plasma: Atomic Scale Simulations

Jamoliddin Razzokov *, Maksudbek Yusupov and Annemie Bogaerts

Research Group PLASMANT, Department of Chemistry, University of Antwerp, Universiteitsplein 1,
B-2610 Antwerp, Belgium; maksudbek.yusupov@uantwerpen.be (M.Y.);
annemie.bogaerts@uantwerpen.be (A.B.)
* Correspondence: jamoliddin.razzokov@uantwerpen.be; Tel.: +32-03-265-2382

Received: 9 May 2018; Accepted: 6 June 2018; Published: 8 June 2018

Abstract: Cold atmospheric plasma (CAP) has shown its potential in biomedical applications, such as wound healing, cancer treatment and bacterial disinfection. Recent experiments have provided evidence that CAP can also enhance the intracellular uptake of glucose molecules which is important in diabetes therapy. In this respect, it is essential to understand the underlying mechanisms of intracellular glucose uptake induced by CAP, which is still unclear. Hence, in this study we try to elucidate the possible mechanism of glucose uptake by cells by performing computer simulations. Specifically, we study the transport of glucose molecules through native and oxidized membranes. Our simulation results show that the free energy barrier for the permeation of glucose molecules across the membrane decreases upon increasing the degree of oxidized lipids in the membrane. This indicates that the glucose permeation rate into cells increases when the CAP oxidation level in the cell membrane is increased.

Keywords: cold atmospheric plasma; reactive oxygen and nitrogen species; glucose uptake; molecular dynamics; permeation free energy

1. Introduction

Diabetes is a chronic disease related to an abnormal increase of the glucose level in the blood. Approximately 75% of glucose in the body is consumed by skeletal muscle cells that are stimulated by insulin [1,2]. Failure of glucose uptake leads to the presence of a high level of sugar in the blood. This is due to one of two mechanisms or a combination of both, i.e., (a) disproportional production of insulin or (b) inadequate sensitivity of cells to insulin [3–5]. Most of the currently available pharmaceuticals are inefficient in the treatment of diabetes, thus alternative insulin independent healing methods need to be found in order to increase glucose uptake by muscle cells.

Recently, Kumar et al. investigated the impact of cold atmospheric plasma (CAP) on the regulation of glucose homeostasis [6]. The experimental results showed that the glucose uptake is significantly enhanced in skeletal muscle cells after plasma treatment, exhibiting the beneficial effects of CAP. Furthermore, higher levels of intracellular Ca^{2+} and reactive oxygen species (ROS) in CAP treated cells were observed. An increase in intracellular ROS and Ca^{2+} ions helps to increase the glucose uptake by skeletal muscle cells [7,8]. Increases in the Ca^{2+} ion concentration as well as the uptake of middle-sized, membrane-impermeable molecules after direct CAP exposure were also observed in [9]. The authors attributed these effects to an increase in cell permeability caused by CAP-generated electric stimulation and the delivery of OH radicals into cells. Moreover, Vijayarangan et al. investigated CAP parameters and conditions for drug delivery across HeLa cells [10]. They determined the efficient treatment time (i.e., low toxicity) and the range of frequencies with an optimal number of pulses as key parameters for the cell membrane permeability [10]. In addition, Leduc et al. determined the maximum radius for macromolecules that are capable of ingressing into HeLa cells, applying a specifically designed CAP source [11]. Hence, an increase in the cell membrane permeability might play an important role in the delivery of the abovementioned species into the cell. The underlying mechanisms, however, still remain unclear, and need more thorough investigation. Computer simulations might be a useful tool to gain insight into the atomic level processes.

In the context of plasma medicine, we have already performed several computational studies, by means of molecular dynamics (MD) simulations, using a phospholipid bilayer (PLB) as a model system for the plasma membrane. Specifically, we studied the effect of lipid oxidation on phosphatidylserine translocation across the plasma membrane, which plays a vital role in apoptosis signaling [12]. Furthermore, we investigated the ROS oxidation of the head groups and lipid tails in the membrane [13], the permeation of ROS across oxidized and non-oxidized membranes, including the synergistic effect of plasma oxidation and electric field [14], and the hampering effect of cholesterol [15,16]. In general, our investigations showed that oxidation of the membrane leads to an increase in its fluidity and permeability to ROS, thereby affecting the abovementioned processes.

In this study, we investigate glucose translocation across native and oxidized membranes in order to provide a possible explanation for the mechanism of glucose uptake observed in previous experiments. In particular, we perform MD simulations to calculate the free energy barriers for glucose transport through oxidized and non-oxidized membranes. Comparison of the latter shows the effect of CAP oxidation on the glucose transport across the membrane, which might explain the increased level of glucose uptake after the plasma treatment of cells, as observed experimentally.

2. Computational Details

2.1. Simulation Setup

We performed MD simulations in order to study the glucose translocation across both intact and oxidized PLBs. The PLB is considered to be a simple model system that represents the eukaryotic cell membrane, since it determines the thickness of the bilayer. A schematic representation of the intact PLB is given in Figure 1a. It consists of 128 phospholipids (PLs), covered by 6000 water molecules, organized in two lamellae (i.e., 64 PLs, with a corresponding water layer at the top, and 64 at the bottom). As the PL molecule, we used 1,2-dioleoyl-sn-glycero-3-phosphocholine (DOPC), as depicted in Figure 1b. Glucose molecules were randomly placed in the xy-plane of the upper side of the PLB, i.e., in the water phase, as well as in the lipid tail region, see Figure 1a (and see below for more details).

Simulations were carried out using the GROMACS package (version 5.1) [17], by applying the GROMOS (43A1-S3) force field [18]. To study the effect of oxidized PLs, we used aldehyde oxidation products (see Figure 1b, DOPC–ALD), which are one of the major oxidation products [19]. Note that CAP yields a cocktail of reactive species and thus, could possibly form a range of products upon oxidation of the PLB, but the formation of aldehyde groups (i.e., DOPC–ALD) was prominently observed in CAP-treated vesicles [13]. These oxidation products were also used in our recent simulation studies on various properties of the PLB [13–15]. The force field parameters for the aldehyde groups in the oxidized PLs were obtained from [20] and the parameters of glucose were based on [21]. The Packmol package was employed to create initial configurations of the intact and oxidized PLB systems [22]. Two aldehyde oxidation products were created from the non-oxidized (i.e., native) PLBs containing 128 PLs, by replacing 32 and 64 DOPC molecules with DOPC–ALD, corresponding to concentrations of 25% and 50%, respectively. The presence of 25% and 50% DOPC–ALD in the oxidized PLB does not necessarily correspond to experimental oxidation levels, but performing the calculations for a lower degree of oxidation would require excessive calculation times to give the same qualitative conclusions. Hence, this oxidation degree was high enough to observe the effect of oxidation within an acceptable calculation time, but low enough so that pore formation did not occur within the simulated time scale [15]. Thus, in total, three model systems were studied in our simulations, i.e., native (0%) as well as aldehyde oxidized (25% and 50%) PLBs. For each system, we created four different structures (e.g., four native PLBs) extracted from the last 40 ns trajectory of the 200 ns equilibration run, with a time interval of 10 ns, in order to obtain an average free energy profile (FEP) of glucose transition across each system (see next section). All structures (in total 12, including the native PLB) are initially optimized using the steepest descent algorithm and then equilibrated for 200 ns by so-called NPT simulations (i.e., at a constant number of particles (N), pressure (P) and temperature (T)), at 310 K and 1 bar, employing the Nose–Hoover thermostat [23] with a coupling constant of 0.2 ps as well as the semi-isotropic Parrinello–Rahman barostat [24] with a compressibility and coupling constant of 4.5×10^{-5} bar^{-1} and 0.1 ps, respectively. For the non-bonded interactions, a 1.1 nm cut-off was applied. The long range electrostatic interactions were described by the particle mesh Ewald (PME) method [25], using a 1.1 nm cut-off for the real-space interactions in combination with a 0.15 nm spaced grid for the reciprocal–space interactions. Subsequently, a series of umbrella sampling (US) simulations [26] were run for 20 ns, applying, again, the NPT ensemble (see next section), of which the last 10 ns was used for further analysis. In all simulations, a time step of 2 fs was used. Periodic boundary conditions were applied in all three directions.

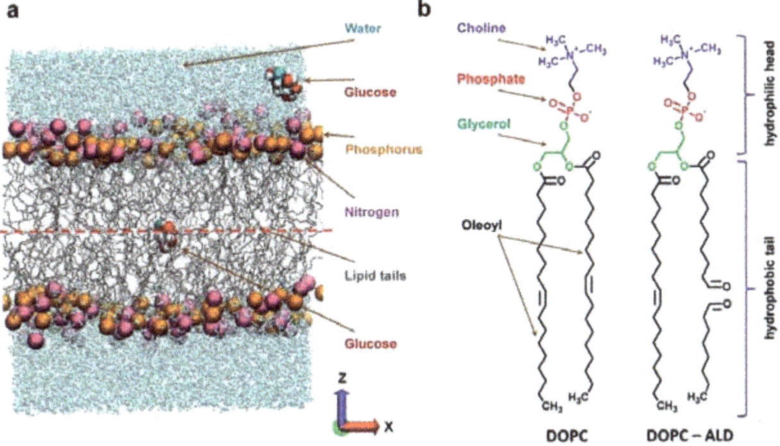

Figure 1. (**a**) Intact or native 1,2-dioleoyl-sn-glycero-3-phosphocholine (DOPC) phospholipid bilayer (PLB), together with two glucose molecules in the water and lipid tail regions. For the sake of clarity, the N and P atoms of DOPC are shown with larger beads. The bilayer center is indicated by the red dashed line; (**b**) Schematic illustration of native (DOPC) and oxidized (DOPC–ALD) PLs. The head group consists of choline, phosphate and glycerol, whereas the lipid tails are two fatty acid chains.

2.2. Umbrella Sampling (US) Simulations

We performed US simulations in order to determine the free energy profiles (FEPs) of glucose translocation across the native, as well as the oxidized, PLBs. In total, we obtained 12 FEPs for all model systems (i.e., four for the native, four for the 25% oxidized, and four for the 50% oxidized PLBs, see previous section) and averaging was performed over four FEPs for each system. For the calculation of each energy profile, we extracted 32 to 36 windows (36 for the native, 34 for the 25% and 32 for the 50% oxidized PLB, due to a decrease of the bilayer thickness) along the z-axis, which were separated by 0.1 nm. These windows were obtained by pulling glucose molecules against the z-axis direction (see Figure 1a) for 500 ps, using a harmonic bias between these glucose molecules and the center of mass of the PLB, with a force constant of 2000 kJ·mol^{-1}·nm^{-2} and a pulling rate of 0.01 nm·ps^{-1}. Note that, in principle, slow pulling rates can be used in the pulling simulations. However, dragging the glucose from water phase into the center of the PLB requires a long computation time. Higher pulling rates can solve this issue, but they can lead to significant disturbances in the PLB. Thus, by using an appropriate (standard) pulling rate, we performed short pulling simulations to save computation time with minimum perturbations on the PLB. One of the glucose molecules (i.e., the glucose in the upper water phase, see Figure 1a) was pulled until it reached the center of mass of the PLB, while the second one moved towards the lower water phase from the center of the bilayer. Thus, each US simulation involved two glucose molecules. In this way, we saved computational resources, thereby increasing the number of sampling points. Note that these two glucose molecules were separated from each other at least by 3 nm in the z direction; hence, there was no interaction between these two molecules. In principle, we could have used three glucose molecules in each US simulation. However, in order to cause minimal disturbance to the PLB system, we chose two glucose molecules instead of three. It should be mentioned that, in reality, adsorption or chemical reaction of glucose might take place in the PLB. However, these processes cannot be studied with conventional non-reactive MD simulations, due to the limitations in time and reactivity. Nevertheless, the US simulations can predict how often the glucose transport occurs across the PLB before and after oxidation, through calculation of the FEPs.

As mentioned above, we extracted 32 to 36 US windows from our 500 ps pulling simulation. Hence, 32 to 36 US simulations were performed to construct a single FEP. Each US simulation lasted 20 ns, and the last 10 ns were used for analysis, i.e., to collect the US histograms and calculate the FEPs. Note that the pulling simulation trajectory was used only to extract windows/frames for the further US simulation to obtain the FEPs. During the pulling simulations, the glucose dragged water molecules with it into the hydrophobic core of the bilayer, but these molecules escaped from the hydrophobic core within the initial 10 ns of US simulation. Thus, the last 10 ns of the US simulation was an adequate time for calculating the FEPs, as there were not any water defects or hydration layers in the hydrophobic part of the membrane. In each US simulation, the glucose molecules were able to freely travel in the xy-plane, but their movement in the z-direction was restricted by applying a harmonic bias with a spring constant of 2000 $kJ \cdot mol^{-1} \cdot nm^{-2}$.

The FEPs were constructed using a periodic version of the weighted histogram analysis method (WHAM) [27], as implemented in GROMACS. We analyzed our simulation systems to identify underestimated possible "hidden barriers" based on previous literature [28]. We did not define any hidden barriers because in our US simulations, the data was sufficiently sampled. Indeed, in the four model systems used, the glucose was randomly positioned in the xy-plane to escape trapped metastable states and to allow estimation of the error bars that are associated with choosing the initial model systems. As noted above, the final energy profiles were obtained by averaging four independently-built FEPs for each system which differed from one another based on their starting structure to allow for some statistical variation. The uncertainties associated with the FEPs were obtained by calculating the standard deviations between these four FEPs for each system. Thus, in total, 410 US simulations were performed for the calculation of the FEPs.

3. Results and Discussion

The US MD simulations allowed us to elucidate the glucose translocation across the native and oxidized PLBs which gave insight into the possible mechanism of glucose ingress triggered by plasma oxidation of the cell membrane. Figure 2 illustrates the symmetric FEPs of glucose transfer across native (0%) as well as oxidized (25% and 50%) PLBs.

It is clear that, in all three cases, the ΔG started to drop when glucose entered the hydrophilic head group region from the water phase (see Figure 1b). This means that the head group of the PLB is energetically the most favourable region, showing the minimum energy for glucose transfer. Moreover, this energy minimum slightly shifted towards the centre of the bilayer in the case of oxidized PLBs. We know from our previous studies that when lipid tails are oxidized, this eventually leads to a drop of the bilayer thickness [13,14]. In other words, the shortened tail and aldehyde products move towards the water phase, thereby increasing the area and fluidity of the PLB. Consequently, this results in a decrease in PLB thickness, e.g., the native DOPC bilayer has a thickness of around 3.89 nm, whereas, in the case of a 50% DOPC–ALD bilayer, the thickness decreases to 3.33 nm [14]. Therefore, the free energy minima in Figure 2 are found at around $|z| = 1.9$ nm and $|z| = 1.76$ nm for native and 50% oxidized PLBs, respectively.

Continuing the motion of glucose towards the hydrophobic tail region leads to an increase of the free energy for translocation, representing the role of the membrane as a permeation barrier. In case of the native PLB, the free energy barrier for the permeation of glucose was 51 ± 3 kJ/mol, which is within the range of experimental results [29,30]. On the other hand, for the 25% and 50% oxidized PLBs, this barrier for the translocation of glucose decreased to values of 37 ± 4 kJ/mol and 28 ± 4 kJ/mol, respectively (see Figure 2). Hence, the obtained results show that the free energy barrier for the transport of glucose molecules across the PLB decreases when the oxidation degree is increased. This, in turn, leads to an increase in the probability of glucose permeation to the cell interior. The obtained simulation results can be correlated with the experimental observations [6,9] as the plasma treatment of cells most probably gives rise to oxidation of the cell membrane, thereby increasing the glucose (or other middle-sized molecules, as well as Ca^{2+}) translocation rate.

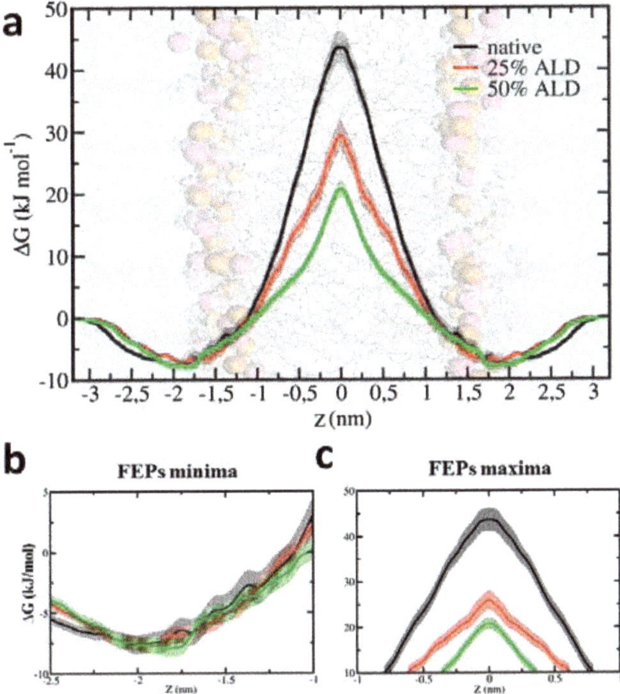

Figure 2. (**a**) Symmetric free energy profiles for the translocation of glucose across the native and oxidized PLBs. A PLB is schematically illustrated in the background, to indicate the position of the water layer, the head groups and the lipid tails. For clarity, the zoomed extrema of the profiles are shown in (**b**,**c**). Errors associated with the umbrella sampling (US) calculations are depicted in pale color.

Note that our simulations only provide one possible explanation for the increased level of glucose [6] in cells after plasma treatment, while other mechanisms might play a role as well [9–11]. Therefore, further investigations should be performed to obtain a more complete picture of the plasma effect on the membrane permeability. This might be achieved, for instance, by silencing the glucose transporter proteins (or GLUTs) and measuring the concentration of intracellular glucose molecules after CAP treatment. The latter would clearly reveal the role of cell membrane permeability (induced by CAP oxidation of lipids) in transporting glucose molecules.

4. Conclusions

We performed MD simulations in order to understand the possible mechanisms of cellular glucose uptake induced by CAP treatment. The obtained free energy profiles of glucose across native and oxidized membranes revealed that the plasma induced oxidation of the membrane lipids decreases the barrier for translocation of glucose across the membrane. This, in turn, might possibly explain the increased concentration of glucose observed by experiments using CAP. Hence, this computational study provides an atomic level insight into the possible process of glucose permeation through the membrane.

Author Contributions: Formal analysis, J.R.; Methodology, J.R. and M.Y.; Visualization, J.R. and M.Y.; Writing—Original Draft Preparation, J.R.; Writing—Review & Editing, M.Y. and A.B.; Supervision, A.B.

Funding: This research was funded by the Research Foundation—Flanders (FWO), grant No. 1200216N.

Acknowledgments: The computational work was carried out using the Turing HPC infrastructure at the CalcUA core facility of the Universiteit Antwerpen, a division of the Flemish Supercomputer Center VSC, funded by the Hercules Foundation, the Flemish Government (department EWI), and the Universiteit Antwerpen.

Conflicts of Interest: The authors declare no conflict of interest.

References

1. Khan, A.; Pessin, J. Insulin regulation of glucose uptake: A complex interplay of intracellular signalling pathways. *Diabetologia* **2002**, *45*, 1475–1483. [PubMed]
2. Saltiel, A.R.; Kahn, C.R. Insulin signalling and the regulation of glucose and lipid metabolism. *Nature* **2001**, *414*, 799. [CrossRef] [PubMed]
3. Kahn, S. The relative contributions of insulin resistance and beta-cell dysfunction to the pathophysiology of type 2 diabetes. *Diabetologia* **2003**, *46*, 3–19. [CrossRef] [PubMed]
4. Cavaghan, M.K.; Ehrmann, D.A.; Polonsky, K.S. Interactions between insulin resistance and insulin secretion in the development of glucose intolerance. *J. Clin. Investig.* **2000**, *106*, 329–333. [CrossRef] [PubMed]
5. Stumvoll, M.; Goldstein, B.J.; van Haeften, T.W. Type 2 diabetes: Principles of pathogenesis and therapy. *Lancet* **2005**, *365*, 1333–1346. [CrossRef]
6. Kumar, N.; Shaw, P.; Razzokov, J.; Yusupov, M.; Attri, P.; Uhm, H.S.; Choi, E.H.; Bogaerts, A. Enhancement of cellular glucose uptake by reactive species: A promising approach for diabetes therapy. *RSC Adv.* **2018**, *8*, 9887–9894. [CrossRef]
7. Fu, A.; Eberhard, C.E.; Screaton, R.A. Role of AMPK in pancreatic beta cell function. *Mol. Cell. Endocrinol.* **2013**, *366*, 127–134. [CrossRef] [PubMed]
8. Higaki, Y.; Mikami, T.; Fujii, N.; Hirshman, M.F.; Koyama, K.; Seino, T.; Tanaka, K.; Goodyear, L.J. Oxidative stress stimulates skeletal muscle glucose uptake through a phosphatidylinositol 3-kinase-dependent pathway. *Am. J. Physiol.-Endocrinol. Metab.* **2008**, *294*, E889–E897. [CrossRef] [PubMed]
9. Sasaki, S.; Hokari, Y.; Kumada, A.; Kanzaki, M.; Kaneko, T. Direct plasma stimuli including electrostimulation and OH radical induce transient increase in intracellular Ca^{2+} and uptake of a middle-size membrane-impermeable molecule. *Plasma Process. Polym.* **2018**, *15*, 1700077. [CrossRef]
10. Vijayarangan, V.; Delalande, A.; Dozias, S.; Pouvesle, J.-M.; Pichon, C.; Robert, E. Cold atmospheric plasma parameters investigation for efficient drug delivery in HeLa cells. *IEEE Trans. Radiat. Plasma Med. Sci.* **2018**, *2*, 109–115. [CrossRef]
11. Leduc, M.; Guay, D.; Leask, R.; Coulombe, S. Cell permeabilization using a non-thermal plasma. *New J. Phys.* **2009**, *11*, 115021. [CrossRef]
12. Razzokov, J.; Yusupov, M.; Vanuytsel, S.; Neyts, E.C.; Bogaerts, A. Phosphatidylserine flip-flop induced by oxidation of the plasma membrane: A better insight by atomic scale modeling. *Plasma Process. Polym.* **2017**, *14*. [CrossRef]
13. Yusupov, M.; Wende, K.; Kupsch, S.; Neyts, E.; Reuter, S.; Bogaerts, A. Effect of head group and lipid tail oxidation in the cell membrane revealed through integrated simulations and experiments. *Sci. Rep.* **2017**, *7*, 5761. [CrossRef] [PubMed]

14. Yusupov, M.; van der Paal, J.; Neyts, E.; Bogaerts, A. Synergistic effect of electric field and lipid oxidation on the permeability of cell membranes. *Biochim. Biophys. Acta* **2017**, *1861*, 839–847. [CrossRef] [PubMed]
15. Van der Paal, J.; Neyts, E.C.; Verlackt, C.C.; Bogaerts, A. Effect of lipid peroxidation on membrane permeability of cancer and normal cells subjected to oxidative stress. *Chem. Sci.* **2016**, *7*, 489–498. [CrossRef] [PubMed]
16. Van der Paal, J.; Verheyen, C.; Neyts, E.C.; Bogaerts, A. Hampering effect of cholesterol on the permeation of reactive oxygen species through phospholipids bilayer: Possible explanation for plasma cancer selectivity. *Sci. Rep.* **2017**, *7*, 39526. [CrossRef] [PubMed]
17. Van der Spoel, D.; Lindahl, E.; Hess, B.; Groenhof, G.; Mark, A.E.; Berendsen, H.J. GROMACS: Fast, flexible, and free. *J. Comput. Chem.* **2005**, *26*, 1701–1718. [CrossRef] [PubMed]
18. Chiu, S.-W.; Pandit, S.A.; Scott, H.; Jakobsson, E. An improved united atom force field for simulation of mixed lipid bilayers. *J. Phys. Chem. B* **2009**, *113*, 2748–2763. [CrossRef] [PubMed]
19. Reis, A.; Domingues, M.; Amado, F.M.; Ferrer-Correia, A.; Domingues, P. Separation of peroxidation products of diacyl-phosphatidylcholines by reversed-phase liquid chromatography–mass spectrometry. *Biomed. Chromatogr.* **2005**, *19*, 129–137. [CrossRef] [PubMed]
20. Wong-Ekkabut, J.; Xu, Z.; Triampo, W.; Tang, I.-M.; Tieleman, D.P.; Monticelli, L. Effect of lipid peroxidation on the properties of lipid bilayers: A molecular dynamics study. *Biophys. J.* **2007**, *93*, 4225–4236. [CrossRef] [PubMed]
21. Geballe, M.T.; Skillman, A.G.; Nicholls, A.; Guthrie, J.P.; Taylor, P.J. The SAMPL2 blind prediction challenge: Introduction and overview. *J. Comput.-Aided Mol. Des.* **2010**, *24*, 259–279. [CrossRef] [PubMed]
22. Martínez, L.; Andrade, R.; Birgin, E.G.; Martínez, J.M. PACKMOL: A package for building initial configurations for molecular dynamics simulations. *J. Comput. Chem.* **2009**, *30*, 2157–2164. [CrossRef] [PubMed]
23. Hoover, W.G. Canonical dynamics: Equilibrium phase-space distributions. *Phys. Rev. A* **1985**, *31*, 1695. [CrossRef]
24. Parrinello, M.; Rahman, A. Polymorphic transitions in single crystals: A new molecular dynamics method. *J. Appl. Phys.* **1981**, *52*, 7182–7190. [CrossRef]
25. Essmann, U.; Perera, L.; Berkowitz, M.L.; Darden, T.; Lee, H.; Pedersen, L.G. A smooth particle mesh Ewald method. *J. Chem. Phys.* **1995**, *103*, 8577–8593. [CrossRef]
26. Kästner, J. Umbrella sampling. *Wiley Interdiscip. Rev. Comput. Mol. Sci.* **2011**, *1*, 932–942. [CrossRef]
27. Kumar, S.; Rosenberg, J.M.; Bouzida, D.; Swendsen, R.H.; Kollman, P.A. The weighted histogram analysis method for free-energy calculations on biomolecules. I. The method. *J. Comput. Chem.* **1992**, *13*, 1011–1021. [CrossRef]
28. Neale, C.; Pomès, R. Sampling errors in free energy simulations of small molecules in lipid bilayers. *Biochim. Biophys. Acta* **2016**, *1858*, 2539–2548. [CrossRef] [PubMed]
29. Bresseleers, G.J.; Goderis, H.L.; Tobback, P.P. Measurement of the glucose permeation rate across phospholipid bilayers using small unilamellar vesicles Effect of membrane composition and temperature. *Biophys. Acta* **1984**, *772*, 374–382. [CrossRef]
30. Sweet, C.; Zull, J. Activation of glucose diffusion from egg lecithin liquid crystals by serum albumin. *Biophys. Acta* **1969**, *173*, 94–103. [CrossRef]

Review

New Hopes for Plasma-Based Cancer Treatment

Hiromasa Tanaka [1,*], Masaaki Mizuno [2], Kenji Ishikawa [3], Shinya Toyokuni [4],
Hiroaki Kajiyama [5], Fumitaka Kikkawa [5] and Masaru Hori [1]

[1]   Institute of Innovation for Future Society, Nagoya University, Furo-cho, Chikusa-ku,
      Nagoya 464-8603, Japan; hori@nuee.nagoya-u.ac.jp
[2]   Center for Advanced Medicine and Clinical Research, Nagoya University Hospital, Tsurumai-cho 65,
      Showa-ku, Nagoya 466-8550, Japan; mmizuno@med.nagoya-u.ac.jp
[3]   Plasma Nanotechnology Research Center, Nagoya University, Furo-cho, Chikusa-ku,
      Nagoya 464-8603, Japan; ishikawa@plasma.engg.nagoya-u.ac.jp
[4]   Department of Pathology and Biological Responses, Nagoya University Graduate School of Medicine,
      Tsurumai-cho 65, Showa-ku, Nagoya 466-8550, Japan; toyokuni@med.nagoya-u.ac.jp
[5]   Department of Obstetrics and Gynecology, Nagoya University Graduate School of Medicine,
      Tsurumai-cho 65, Showa-ku, Nagoya 466-8550, Japan; kajiyama@med.nagoya-u.ac.jp (H.K.);
      kikkawaf@med.nagoya-u.ac.jp (F.K.)
*   Correspondence: htanaka@plasma.engg.nagoya-u.ac.jp; Tel.: +81-52-788-6230

Received: 30 July 2018; Accepted: 16 August 2018; Published: 18 August 2018

Abstract: Non-thermal plasma represents a novel approach in cancer treatment. Both direct and indirect plasma treatments are available, with clinical trials of direct plasma treatment in progress. Indirect treatments involve chemotherapy (i.e., plasma-activated medium) and immunotherapy. Recent studies suggest that integrated plasma treatments could be an extremely effective approach to cancer therapy.

Keywords: plasma cancer treatment; plasma-activated medium (PAM); plasma-assisted immunotherapy

1. Introduction

Plasma-based cancer treatments represent a critical area in the field of plasma medicine [1–3]. Some pioneering in vitro [4] and in vivo [5] works have shown that non-thermal plasma exerts anti-tumor effects. Currently, two options for plasma cancer treatment are available: direct and indirect (Figure 1). Clinical trials of cancer treatments using non-thermal plasma (direct plasma treatments) are ongoing in Germany [6] and the USA [7]. Two different types of indirect plasma treatment have been proposed: plasma-assisted immunotherapy [8] and plasma-activated medium (PAM) [9,10]. Plasma is also considered as an adjuvant therapy, and the three main options are plasma in combination with chemotherapy [11,12], plasma to modulate tumor microenvironment [13,14], and plasma in association with electrotherapy [15]. These approaches have dramatically broadened the ways of using non-thermal plasma for treating cancers and other diseases.

Currently, two options for plasma cancer treatment are available: direct and indirect. Clinical trials of direct plasma treatments are ongoing. Indirect treatments include plasma-assisted cancer immunotherapy and plasma-activated medium (PAM) therapy.

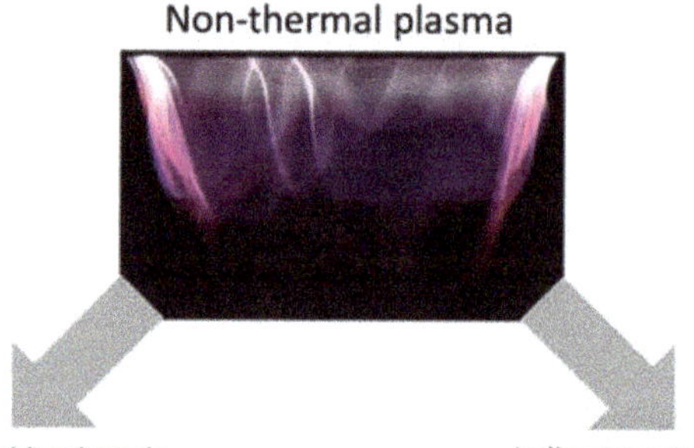

Figure 1. Two types of plasma-based cancer treatments.

2. Direct Treatments

Direct plasma treatments are the most straightforward. A variety of plasma sources have been developed for medical applications such as cancer treatment [16–20]. Metelmann et al. used a plasma source known as kINPen MED to treat advanced head and neck carcinoma ulcerations and patients in the final stages of disease [6,21]. Plasma treatment reduced both pain and odor. Only a few myeloid cells were present in tumor tissue of patients that received frequent plasma treatment, whereas numerous myeloid cells were found in tissue sections of patients that did not receive plasma treatment. Canady et al. treated liver cancer using a Canady Helios Cold Plasma Scalpel to remove cancerous tissue without damaging the blood supply to the remaining liver.

Since plasma needle was used for treatment of culture cells [22], various plasma sources have been developed for cancer treatment. The plasma jet and dielectric barrier discharge (DBD) have been developed and widely used in plasma cancer treatment. A pulsed DBD with microsecond pulses was used for treatment of xenograft model mouse of human glioblastoma cells [23]. Recently, nanosecond-pulsed plasma have been developed as potential tools in cancer treatment [24,25].

3. Indirect Treatment: Plasma-Assisted Cancer Immunotherapy

Several researchers have proposed the use of plasmas as immune modulators for treating cancer. The number of cells in the human body is estimated at about 40 trillion, and a small portion of these cells acquire mutations and become cancerous every day. However, the immune system typically removes mutated cells. It is only when cancerous cells avoid the immune system that cancerous disease develops. Recently, a variety of anti-cancer therapies designed to modulate the immune system have been developed [26–28]. These approaches include the use of cytokines, cell-based therapies, and immune checkpoint blockade. For example, the US Food and Drug Administration (FDA) approved the first cellular immunotherapy (sipuleucel-T) for prostate cancer in 2010 [29]. The FDA also approved the anti-PD1 monoclonal antibody, nivolumab, for adjuvant treatment of patients with melanoma involving the lymph nodes and patients with metastatic disease who have undergone complete resection [30].

A better understanding of the interactions between cancer cells and the immune system has increased interest in immunotherapies over the last decade [31,32]. Radiation and some chemotherapeutic drugs increase immunogenicity by triggering immunogenic cell death (ICD). Damaged or stressed cancer cells present "danger signals" known as damage-associated molecular pattern (DAMP) molecules. High-mobility group box 1, ATP, and calreticulin (CRT) are well-known DAMP molecules that are retained inside the cell in the healthy state and released only in response to stress or cell damage. Cancer cells usually induce immunosuppression; however, ICD-associated

DAMP molecules can reactivate anti-cancer immunity by triggering dendritic cell maturation and antigen presentation.

Several recent studies have suggested that non-thermal plasma treatment induces ICD and stimulates macrophages [25,33–35]. Non-thermal plasma treatment was shown to stimulate extracellular ATP secretion and enhance cell death via ICD-mediated macrophage stimulation. Plasma-generated reactive oxygen species (ROS) are major effectors of ICD. Non-thermal plasma elicits surface exposure of CRT, and N-acetyl cysteine, which is an ROS scavenger, reduces the externalization of CRT. These results suggest that intracellular ROS are responsible for plasma-induced CRT production. Tumor necrosis factor–alpha released from plasma-activated macrophages induces tumor cell death [36].

In the future, non-thermal plasma will be used to induce ICD in tumors to help dendritic cells find, eat, and present cancer cell antigen to elicit robust T cell immune responses [37]. It was shown that naïve T helper cells were less sensitive toward non-thermal plasma treatment, suggesting that plasma could be used as a tool to redox-control T cell phenotypes in cancer immunology [38]. Flow cytometric technique for microparticle characterization was established, and the number and size of microparticles released were shown to be modulated in THP-1 monocytes, polymorphonuclear leukocytes (PMN), and peripheral blood mononuclear cells (PBMC) after plasma exposure [39]. Interestingly, abscopal effects of non-thermal plasma treatment on tumor growth were observed, suggesting that plasma activated innate immune response [24].

4. Indirect Treatment: PAM

PAM has been proposed as a type of cancer chemotherapy. Various in vitro experiments have demonstrated that PAM exerts anti-tumor effects on many kinds of cancer cells [9,10,40,41]. In most cases, PAM induces intracellular ROS production and subsequent apoptosis of cancer cells. The mechanism through which PAM induces the apoptosis of cancer cells depends on the cell type [42,43]. In glioblastoma cells, down-regulation of survival and proliferation signaling networks plays a critical role in PAM-induced apoptosis [9,42,44]. Aquaporins, which transport hydrogen peroxide into the plasma membrane, are also key factors in apoptosis induction [45]. Many in vivo experiments have also demonstrated the anti-tumor effects of PAM against a variety of cancers. In a xenograft mouse model, PAM inhibited the growth of ovarian and pancreatic cancer tumor cells [10,41]. Intraperitoneal injection of PAM/plasma-activated Ringer's lactate (PAL) inhibited the metastasis of ovarian, gastric, and pancreatic cancer tumors in disease model mouse experiments examining peritoneal metastasis [46–48]. However, in apoptosis induced by PAL, less ROS are produced in comparison with PAM [49], suggesting that components generated in PAM control the redox balance.

5. Conclusions

Two major options are available for plasma-based cancer therapies: direct and indirect treatments. Direct plasma treatment methods have already been introduced clinically, whereas indirect plasma treatment methods such as plasma-assisted cancer immunotherapy and PAM therapy are new approaches currently under study. In the future, the overall survival of cancer patients could be significantly improved by combining direct and indirect plasma treatments.

Funding: This work was funded in part by Grants-in-Aid for Scientific Research on Innovative Areas "Plasma Medical Innovation" (No. 24108002 and No. 24108008), a Grant-in-Aid for Young Scientists (A) (No. 15H05430), a Grant-in-Aid for Challenging Exploratory Research Grant (No. 15K13390), and Grant-in-Aid for Scientific Research (C) (No. 18K03599) from the Ministry of Education, Culture, Sports, Science and Technology of Japan.

Conflicts of Interest: The authors declare no conflicts of interest.

References

1. Laroussi, M. From killing bacteria to destroying cancer cells: 20 years of plasma medicine. *Plasma Process. Polym.* **2014**, *11*, 1138–1141. [CrossRef]
2. Tanaka, H.; Ishikawa, K.; Mizuno, M.; Toyokuni, S.; Kajiyama, H.; Kikkawa, F.; Metelmann, H.R.; Hori, M. State of the art in medical applications using non-thermal atmospheric pressure plasma. *Rev. Mod. Plasma Phys.* **2017**, *1*, 3. [CrossRef]
3. Schlegel, J.; Köritzer, J.; Boxhammer, V. Plasma in cancer treatment. *Clin. Plasma Med.* **2013**, *1*, 2–7. [CrossRef]
4. Stoffels, E.; Kieft, I.E.; Sladek, R.E.J.; van den Bedem, L.J.M.; van der Laan, E.P.; Steinbuch, M. Plasma needle for in vivo medical treatment: Recent developments and perspectives. *Plasma Sources Sci. Technol.* **2006**, *15*, S169–S180. [CrossRef]
5. Vandamme, M.; Robert, E.; Lerondel, S.; Sarron, V.; Ries, D.; Dozias, S.; Sobilo, J.; Gosset, D.; Kieda, C.; Legrain, B.; et al. Ros implication in a new antitumor strategy based on non-thermal plasma. *Int. J. Cancer* **2012**, *130*, 2185–2194. [CrossRef] [PubMed]
6. Metelmann, H.R.; Nedrelow, D.S.; Seebauer, C.; Schuster, M.; von Woedtke, T.; Weltmann, K.D.; Kindler, S.; Metelmann, P.H.; Finkelstein, S.E.; Von Hoff, D.D.; et al. Head and neck cancer treatment and physical plasma. *Clin. Plasma Med.* **2015**, *3*, 17–23. [CrossRef]
7. Shashurin, A.; Scott, D.; Zhuang, T.S.; Canady, J.; Beilis, I.I.; Keidar, M. Electric discharge during electrosurgery. *Sci. Rep.* **2015**, *5*, 9946. [CrossRef] [PubMed]
8. Miller, V.; Lin, A.; Fridman, G.; Dobrynin, D.; Fridman, A. Plasma stimulation of migration of macrophages. *Plasma Process. Polym.* **2014**, *11*, 1193–1197. [CrossRef]
9. Tanaka, H.; Mizuno, M.; Ishikawa, K.; Nakamura, K.; Kajiyama, H.; Kano, H.; Kikkawa, F.; Hori, M. Plasma-activated medium selectively kills glioblastoma brain tumor cells by down-regulating a survival signaling molecule, akt kinase. *Plasma Med.* **2013**, *1*, 265–277. [CrossRef]
10. Utsumi, F.; Kajiyama, H.; Nakamura, K.; Tanaka, H.; Mizuno, M.; Ishikawa, K.; Kondo, H.; Kano, H.; Hori, M.; Kikkawa, F. Effect of indirect nonequilibrium atmospheric pressure plasma on anti-proliferative activity against chronic chemo-resistant ovarian cancer cells in vitro and in vivo. *PLoS ONE* **2013**, *8*, e81576. [CrossRef] [PubMed]
11. Brulle, L.; Vandamme, M.; Ries, D.; Martel, E.; Robert, E.; Lerondel, S.; Trichet, V.; Richard, S.; Pouvesle, J.M.; Le Pape, A. Effects of a non thermal plasma treatment alone or in combination with gemcitabine in a MIA PaCa2-luc orthotopic pancreatic carcinoma model. *PLoS ONE* **2012**, *7*, e52653. [CrossRef] [PubMed]
12. Koritzer, J.; Boxhammer, V.; Schafer, A.; Shimizu, T.; Klampfl, T.G.; Li, Y.F.; Welz, C.; Schwenk-Zieger, S.; Morfill, G.E.; Zimmermann, J.L.; et al. Restoration of sensitivity in chemo–resistant glioma cells by cold atmospheric plasma. *PLoS ONE* **2013**, *8*, e64498. [CrossRef] [PubMed]
13. Collet, G.; Robert, E.; Lenoir, A.; Vandamme, M.; Darny, T.; Dozias, S.; Kieda, C.; Pouvesle, J.M. Plasma jet-induced tissue oxygenation: Potentialities for new therapeutic strategies. *Plasma Sources Sci. Technol.* **2014**, *23*, 012005. [CrossRef]
14. Kisch, T.; Schleusser, S.; Helmke, A.; Mauss, K.L.; Wenzel, E.T.; Hasemann, B.; Mailaender, P.; Kraemer, R. The repetitive use of non-thermal dielectric barrier discharge plasma boosts cutaneous microcirculatory effects. *Microvasc. Res.* **2016**, *106*, 8–13. [CrossRef] [PubMed]
15. Daeschlein, G.; Scholz, S.; Lutze, S.; Arnold, A.; von Podewils, S.; Kiefer, T.; Tueting, T.; Hardt, O.; Haase, H.; Grisk, O.; et al. Comparison between cold plasma, electrochemotherapy and combined therapy in a melanoma mouse model. *Exp. Dermatol.* **2013**, *22*, 582–586. [CrossRef] [PubMed]
16. Fridman, G.; Friedman, G.; Gutsol, A.; Shekhter, A.B.; Vasilets, V.N.; Fridman, A. Applied plasma medicine. *Plasma Process. Polym.* **2008**, *5*, 503–533. [CrossRef]
17. Weltmann, K.D.; von Woedtke, T. Basic requirements for plasma sources in medicine. *Eur. Phys. J. Appl. Phys.* **2011**, *55*, 13807. [CrossRef]
18. Kong, M.G.; Kroesen, G.; Morfill, G.; Nosenko, T.; Shimizu, T.; van Dijk, J.; Zimmermann, J.L. Plasma medicine: An introductory review. *New J. Phys.* **2009**, *11*, 115012. [CrossRef]
19. Laroussi, M.; Fridman, A.; Satava, R.M. Plasma medicine. *Plasma Process. Polym.* **2008**, *5*, 501–502. [CrossRef]
20. Morfill, G.E.; Kong, M.G.; Zimmermann, J.L. Focus on plasma medicine. *New J. Phys.* **2009**, *11*, 115011. [CrossRef]

21. Metelmann, H.R.; Seebauer, C.; Miller, V.; Fridman, A.; Bauer, G.; Graves, D.B.; Pouvesle, J.M.; Rutkowski, R.; Schuster, M.; Bekeschus, S.; et al. Clinical experience with cold plasma in the treatment of locally advanced head and neck cancer. *Clin. Plasma Med.* **2018**, *9*, 6–13. [CrossRef]

22. Kieft, I.E.; Broers, J.L.V.; Caubet-Hilloutou, V.; Slaaf, D.W.; Ramaekers, F.C.S.; Stoffels, E. Electric discharge plasmas influence attachment of cultured cho k1 cells. *Bioelectromagnetics* **2004**, *25*, 362–368. [CrossRef] [PubMed]

23. Vandamme, M.; Robert, E.; Pesnel, S.; Barbosa, E.; Dozias, S.; Sobilo, J.; Lerondel, S.; Le Pape, A.; Pouvesle, J.M. Antitumor effect of plasma treatment on u87 glioma xenografts: Preliminary results. *Plasma Process. Polym.* **2010**, *7*, 264–273. [CrossRef]

24. Mizuno, K.; Yonetamari, K.; Shirakawa, Y.; Akiyama, T.; Ono, R. Anti-tumor immune response induced by nanosecond pulsed streamer discharge in mice. *J. Phys. D Appl. Phys.* **2017**, *50*, 12LT01. [CrossRef]

25. Lin, A.; Truong, B.; Patel, S.; Kaushik, N.; Choi, E.H.; Fridman, G.; Fridman, A.; Miller, V. Nanosecond-pulsed DBD plasma-generated reactive oxygen species trigger immunogenic cell death in A549 lung carcinoma cells through intracellular oxidative stress. *Int. J. Mol. Sci.* **2017**, *18*, 966. [CrossRef] [PubMed]

26. Farkona, S.; Diamandis, E.P.; Blasutig, I.M. Cancer immunotherapy: The beginning of the end of cancer? *BMC Med.* **2016**, *14*, 73. [CrossRef] [PubMed]

27. Chen, D.S.; Mellman, I. Elements of cancer immunity and the cancer-immune set point. *Nature* **2017**, *541*, 321–330. [CrossRef] [PubMed]

28. Kirkwood, J.M.; Butterfield, L.H.; Tarhini, A.A.; Zarour, H.; Kalinski, P.; Ferrone, S. Immunotherapy of cancer in 2012. *CA Cancer J. Clin.* **2012**, *62*, 309–335. [CrossRef] [PubMed]

29. Topalian, S.L.; Weiner, G.J.; Pardoll, D.M. Cancer immunotherapy comes of age. *J. Clin. Oncol.* **2011**, *29*, 4828–4836. [CrossRef] [PubMed]

30. Okazaki, T.; Chikuma, S.; Iwai, Y.; Fagarasan, S.; Honjo, T. A rheostat for immune responses: The unique properties of PD-1 and their advantages for clinical application. *Nat. Immunol.* **2013**, *14*, 1212–1218. [CrossRef] [PubMed]

31. Radogna, F.; Diederich, M. Stress-induced cellular responses in immunogenic cell death: Implications for cancer immunotherapy. *Biochem. Pharmacol.* **2018**, *153*, 12–23. [CrossRef] [PubMed]

32. Hernandez, C.; Huebener, P.; Schwabe, R.F. Damage-associated molecular patterns in cancer: A double-edged sword. *Oncogene* **2016**, *35*, 5931–5941. [CrossRef] [PubMed]

33. Miller, V.; Lin, A.; Kako, F.; Gabunia, K.; Kelemen, S.; Brettschneider, J.; Fridman, G.; Fridman, A.; Autieri, M. Microsecond-pulsed dielectric barrier discharge plasma stimulation of tissue macrophages for treatment of peripheral vascular disease. *Phys. Plasmas* **2015**, *22*, 122005. [CrossRef] [PubMed]

34. Miller, V.; Lin, A.; Fridman, A. Why target immune cells for plasma treatment of cancer. *Plasma Chem. Plasma Process.* **2016**, *36*, 259–268. [CrossRef]

35. Lin, A.; Truong, B.; Pappas, A.; Kirifides, L.; Oubarri, A.; Chen, S.Y.; Lin, S.J.; Dobrynin, D.; Fridman, G.; Fridman, A.; et al. Uniform nanosecond pulsed dielectric barrier discharge plasma enhances anti-tumor effects by induction of immunogenic cell death in tumors and stimulation of macrophages. *Plasma Process. Polym.* **2015**, *12*, 1392–1399. [CrossRef]

36. Kaushik, N.K.; Kaushik, N.; Min, B.; Choi, K.H.; Hong, Y.J.; Miller, V.; Fridman, A.; Choi, E.H. Cytotoxic macrophage-released tumour necrosis factor-alpha (TNF-alpha) as a killing mechanism for cancer cell death after cold plasma activation. *J. Phys. D Appl. Phys.* **2016**, *49*, 084001. [CrossRef]

37. Bekeschus, S.; Favia, P.; Robert, E.; von Woedtke, T. White paper on plasma for medicine and hygiene: Future in plasma health sciences. *Plasma Process. Polym.* **2018**. [CrossRef]

38. Bekeschus, S.; Rodder, K.; Schmidt, A.; Stope, M.B.; von Woedtke, T.; Miller, V.; Fridman, A.; Weltmann, K.D.; Masur, K.; Metelmann, H.R.; et al. Cold physical plasma selects for specific t helper cell subsets with distinct cells surface markers in a caspase-dependent and NF-kappa B-independent manner. *Plasma Process. Polym.* **2016**, *13*, 1144–1150. [CrossRef]

39. Bekeschus, S.; Moritz, J.; Schmidt, A.; Wende, K. Redox regulation of leukocyte-derived microparticle release and protein content in response to cold physical plasma-derived oxidants. *Clin. Plasma Med.* **2017**, *7*, 24–35. [CrossRef]

40. Torii, K.; Yamada, S.; Nakamura, K.; Tanaka, H.; Kajiyama, H.; Tanahashi, K.; Iwata, N.; Kanda, M.; Kobayashi, D.; Tanaka, C.; et al. Effectiveness of plasma treatment on gastric cancer cells. *Gastric Cancer* **2014**, *18*, 635–643. [CrossRef] [PubMed]

41. Hattori, N.; Yamada, S.; Torii, K.; Takeda, S.; Nakamura, K.; Tanaka, H.; Kajiyama, H.; Kanda, M.; Fujii, T.; Nakayama, G.; et al. Effectiveness of plasma treatment on pancreatic cancer cells. *Int. J. Oncol.* **2015**, *47*, 1655–1662. [CrossRef] [PubMed]

42. Tanaka, H.; Mizuno, M.; Ishikawa, K.; Nakamura, K.; Utsumi, F.; Kajiyama, H.; Kano, H.; Maruyama, S.; Kikkawa, F.; Hori, M. Cell survival and proliferation signaling pathways are downregulated by plasma-activated medium in glioblastoma brain tumor cells. *Plasma Med.* **2014**, *2*, 207–220. [CrossRef]

43. Adachi, T.; Tanaka, H.; Nonomura, S.; Hara, H.; Kondo, S.I.; Hori, M. Plasma-activated medium induces A549 cell injury via a spiral apoptotic cascade involving the mitochondrial-nuclear network. *Free Radical Biol. Med.* **2014**, *79*, 28–44. [CrossRef] [PubMed]

44. Tanaka, H.; Mizuno, M.; Ishikawa, K.; Takeda, K.; Nakamura, K.; Utsumi, F.; Kajiyama, H.; Kano, H.; Okazaki, Y.; Toyokuni, S.; et al. Plasma medical science for cancer therapy: Toward cancer therapy using nonthermal atmospheric pressure plasma. *IEEE Trans. Plasma Sci.* **2014**, *42*, 3760–3764. [CrossRef]

45. Yan, D.Y.; Xiao, H.J.; Zhu, W.; Nourmohammadi, N.; Zhang, L.G.; Bian, K.; Keidar, M. The role of aquaporins in the anti-glioblastoma capacity of the cold plasma-stimulated medium. *J. Phys. D Appl. Phys.* **2017**, *50*, 055401. [CrossRef]

46. Takeda, S.; Yamada, S.; Hattori, N.; Nakamura, K.; Tanaka, H.; Kajiyama, H.; Kanda, M.; Kobayashi, D.; Tanaka, C.; Fujii, T.; et al. Intraperitoneal administration of plasma-activated medium: Proposal of a novel treatment option for peritoneal metastasis from gastric cancer. *Ann. Surg. Oncol.* **2017**, *24*, 1188–1194. [CrossRef] [PubMed]

47. Nakamura, K.; Peng, Y.; Utsumi, F.; Tanaka, H.; Mizuno, M.; Toyokuni, S.; Hori, M.; Kikkawa, F.; Kajiyama, H. Novel intraperitoneal treatment with non-thermal plasma-activated medium inhibits metastatic potential of ovarian cancer cells. *Sci. Rep.* **2017**, *7*, 6085. [CrossRef] [PubMed]

48. Sato, Y.; Yamada, S.; Takeda, S.; Hattori, N.; Nakamura, K.; Tanaka, H.; Mizuno, M.; Hori, M.; Kodera, Y. Effect of plasma-activated lactated ringer's solution on pancreatic cancer cells in vitro and in vivo. *Ann. Surg. Oncol.* **2018**, *25*, 299–307. [CrossRef] [PubMed]

49. Tanaka, H.; Nakamura, K.; Mizuno, M.; Ishikawa, K.; Takeda, K.; Kajiyama, H.; Utsumi, F.; Kikkawa, F.; Hori, M. Non-thermal atmospheric pressure plasma activates lactate in ringer's solution for anti-tumor effects. *Sci. Rep.* **2016**, *6*, 36282. [CrossRef] [PubMed]

MDPI
St. Alban-Anlage 66
4052 Basel
Switzerland
Tel. +41 61 683 77 34
Fax +41 61 302 89 18
www.mdpi.com

MDPI Books Editorial Office
E-mail: books@mdpi.com
www.mdpi.com/books